Harold Clarence Ernst

Infectiousness of Milk

Harold Clarence Ernst

Infectiousness of Milk

ISBN/EAN: 9783337375027

Printed in Europe, USA, Canada, Australia, Japan

Cover: Foto ©berggeist007 / pixelio.de

More available books at **www.hansebooks.com**

INFECTIOUSNESS OF MILK

RESULT OF INVESTIGATIONS MADE FOR THE
TRUSTEES OF THE MASSACHUSETTS
SOCIETY FOR PROMOTING
AGRICULTURE

BOSTON
Published by the Society
1895

The Riverside Press, Cambridge, Mass., U. S. A.
Electrotyped and Printed by H. O. Houghton & Co.

PREFACE.

In the spring of 1887 the trustees of the Massachusetts Society for Promoting Agriculture decided to establish an Experiment Station at "Forest Hills," in Boston, to study the dangers to human life and health arising from the food products of cattle.

Attention had been publicly called to this subject, and the trustees decided that it would be of value to both producer and consumer to have those dangers investigated.

They appointed a special committee to have charge of the work, consisting of Messrs. Thomas Motley, E. F. Bowditch,* Jacob C. Rogers, and Francis H. Appleton.

This committee secured the services of leading specialists to have the direction of the investigations, and this volume contains the report of the work done, and results obtained.

These results were reached in the autumn of 1890, and were reported to the Legislature of 1891, on January 13th, by petition asking for legislation to secure an inspection of cattle in Massachusetts.

It was not until June 11th that a Resolve was passed, but with an appropriation much less than was recommended, and so small as to negative the purpose of the Resolve.

* Deceased.

THE MASSACHUSETTS SOCIETY

FOR

PROMOTING AGRICULTURE.

INFECTIOUSNESS OF MILK.

REPORT

OF WORK DONE UNDER THE AUSPICES OF THE MAS-
SACHUSETTS SOCIETY FOR PROMOTING AGRICULTURE,
UPON THE INFECTIOUSNESS OF MILK FROM TUBER-
CULOUS COWS WITH NO LESION OF THE UDDER.

NOVEMBER, 1894.

*To the Trustees of the Massachusetts Society for Promoting
Agriculture : —*

GENTLEMEN : I have the honor of presenting herewith the
report of the work done under your auspices, upon the question of the infectiousness of the milk from cows affected with
tuberculosis. This report was not completed immediately
upon the close of the experiments, for the reason that the
committee of the trustees in charge did not desire it, and the
delay in its production since I was requested to write it last
fall has been absolutely unavoidable.

My connection with the work was, at first, simply that of
an expert microscopist; but after the first year its entire
direction lay in my hands, with the society's veterinarian,
Dr. Austin Peters, as first assistant. Much of the actual
manipulation was carried on by him and by Dr. Henry Jackson and Langdon Frothingham, D. V. M.

The desire of the committee was to determine whether or
not the milk derived from tuberculous cows might contain
the infectious material of the disease, and in this way become
dangerous when used as an article of food. And this question was of necessity to be divided into two parts: 1st,
whether this infection, if it existed, was confined to cases in
which there was actual tuberculous disease of the udder; and,
2d, whether it might exist in cases in which the udder was

apparently or actually healthy, but the disease existed in other parts of the body.

In regard to the first part of the question, plain common sense showed that the danger of infection was a real one, and, besides this, there existed at the time sufficient experimental data to prove the fact, so that there was very little dispute that, under such circumstances, milk should not be used for food, — certainly in an uncooked condition. Evidence since then in the same direction has constantly accumulated, and now there is hardly a dissenting opinion that milk from cows with tuberculosis of the udder should be condemned for food.

Upon the second point, however, as to whether the milk from cows with tuberculosis, but not of the udder, might be dangerous, there was a great diversity of opinion, and almost no experimental evidence upon which to base what opinion there was. It was in this direction, therefore, that it was especially desirable to obtain evidence, and, after considerable discussion, it was decided that the main line of experiment should be so conducted that this point might be decided. In this, as in everything else, it is to be remembered that one piece of positive evidence obtained under proper conditions is worth many negative results, and it is for this reason that so much value may be attached to the results which have been obtained. These were published in an incomplete form, and have been very widely disseminated, having appeared in full in the "Transactions of the Association of American Physicians," "The American Journal of the Medical Sciences," "The Practitioner," "The New Hampshire State Board of Health Report," "The Bulletin of the Agricultural Experiment Station," "Transactions of the Congrès sur la Tuberculose," the "Centralblatt für Bakteriologie," besides having been largely quoted in many other ways. It is unquestionable that they have had much influence in moulding public opinion in this matter, and at least one direct result of the work has been the inspection of the herds of cattle in New York by the board of health of that state.

The work, then, was undertaken with this special end in

view, to determine *whether the infectious element of tuberculosis ever existed in milk from tuberculous cows whose udders were apparently healthy*, and was prosecuted under the following headings: 1. *A careful and persistent microscopic examination of the milk from such cattle;* 2. *Inoculation experiments with such milk;* 3. *Feeding experiments with the same milk.* In addition to these three main lines of investigation, there was also undertaken: 4. *Similar investigations of the milk supply of Boston;* and, 5. *The gathering of as much evidence as possible from medical men and veterinarians as to cases of probable infection through tuberculous milk that had come under their observation.*

The methods by which these points were observed and the results that were obtained are given below. The experimental farm, where the animals were kept, was at Mattapan, where it was possible to obtain the best hygienic conditions, and where the feeding experiments and post-mortem examinations were conducted, while the remainder of the work was done at the bacteriological laboratory of the Harvard Medical School.

I.

Cover-glass examinations of milk from cows affected with tuberculosis, but, so far as the best veterinary examination could determine, with no disease of the udder.

These examinations were made methodically and continuously for over two years. The milk was collected in sterilized flasks, after the udder had been cleansed as perfectly as possible, and the hands of the operator sterilized. It was taken at once to the laboratory and allowed to stand, carefully protected, over night, and sometimes longer. Different portions were then taken for examination from both the bottom and the top of the fluid. In all cases at least a dozen cover-glasses were used for each examination, and at least fifteen minutes was spent over each cover-glass. The staining employed was invariably the Koch-Ehrlich 24-hour method.

The results of this line of investigation are given below in

TABLE I.

EXAMINATIONS OF MILK FOR BACILLI OF TUBERCULOSIS.

Number and Source.	Date.	Symptoms and physical signs of cow.	Result.	Post-mortem, if any made.
1 Dutton cow.	1887 Dec. 14	T. = 103.5. Emaciation. No signs in udder.	Neg.	Butcher's autopsy = Tuberculosis of both lungs, pleurae, pericardium, and mediastinum.
2 J. C. R.	Dec. 17	Healthy in May–Dec. = + d lymphatics. Dull and — d resps. on r. side cough.	"	None.
3 J. C. R.	"	Healthy in May–Dec. = + d submaxillary glands.	"	"
4 J. C. R.	"	Ibid. dull l. lung.	"	"
5 J. C. R.	"	Ibid. as 3.	"	"
6 J. C. R.	"	Ibid.	"	"
7 J. C. R.	"	Ibid.	"	"
8 J. C. R.	"	Ibid.	"	"
9 F. L. A.	Dec. 14		Pos.	Dr. Peters, Dec. 22. Tuberculosis of lungs, pleura, liver, spleen, peritoneum, ovaries, and mediastinum. Udder healthy.
10 J. C. R.	1888 Jan. 1		No record.	None.
11 J. C. R.	"		"	"
12 Mrs. B.	1887 Dec. 22	+ d sub-maxillary glands. Dull on r. side. Udder healthy.	Neg.	"
13 McLean Asylum.	1888 Jan. 23	Cough at times. Udder indurated.		Milk spoiled and ex. not completed.
14 McLean Asylum.	"	"		
15 McLean Asylum.	"	"		
16 H. A. D.	Feb. 8	Cough. Rough resp. over r. lung. Posterior quarters of udder indurated.	Neg.	None.

TABLE I. (*continued*).

Number and Source.	Date.	Symptoms and physical signs of cow.	Result.	Post-mortem, if any made.
17 Cow A.	Feb. 18	Sub-max. glands + d. Nodulated post. half udder. General appearance good. Milk from post. udder.	Pos.	Jan. 9, 1889. R. lung slightly tuberculous. + d inguinal gland. Udder = no tuberculosis. Lung showed bacilli.
18 Cow A.	March 10	Milk from l. hind teat.	Neg.	" "
19 Cow A.	June 2	Morning's milk.	"	" "
20 Cow A.	Nov. 8	" "	"	" "
21 Cow B.	Feb. 29	8–10 yrs. Pulmonary tub. and probably elsewhere. Emaciation and cough. Udder healthy.	"	July 6, 1889. Gen. tuberculosis of lungs, pleura, liver, peritoneum, spleen, ovaries, uterus, and glands. Udder scirrhous, but no tuberculosis. Sections from lung showed bacilli.
22 Cow B.	March 10	"	"	Sections from udder showed *no* bacilli.
23 Cow B.	April 3	"	"	
24 Cow B.	March 25	"	"	" "
25 Cow B.	June 1, eve.	"	"	" "
26 Cow B.	June 2, morn.	"	"	" "
27 Cow C.	Feb. 29	R. post. quarter. of udder nodulated. No other symptom.	"	None.
28 Cow C.	March 10		"	" "
29 Cow E.	March 28	+ d sub-max. glands. Dull and râles lower r. lung. Little cough.	"	June 21, 1890. Ant. r. lobe tuberculous mass; abscesses upper post. lobe both lungs. *Bacilli present.* Udder few nodules r. post. quarter. *No bacilli.*
30 Cow E.	Nov. 8, morn.		"	" "

TABLE I. (*continued*).

Number and Source.	Date.	Symptoms and physical signs of cow.	Result.	Post-mortem, if any made.
31 Cow H.	March 28	+ sub-max. glands. Probable disease of both lungs. L. post. quarter udder nodulated.	"	April 10, 1889. Few tubercles l. lung, few + d and cheesy glands in mesentery. Few small tubercles in walls of small intestine. Udder slightly fibrous. Bacilli in lungs and gland. *Not* in udder.
32 Cow H.	Nov. 8		"	
33 Cow D.	April 10	+ d sub-max. glands. Probably pleuritic friction lower r. lung.	"	Nov. 21, 1889. + d ing. gland. *No* bacilli. Cheesy nodule in liver, abscesses in both lungs show bacilli. Udder scirrhous. *No* bacilli.
34 Cow D.	Nov. 8		Pos.	" "
35 Cow D.	Dec. 11	From nodulated teat.	Neg.	" "
36 Slocum.	April 19	Probable general tuberc.	"	None.
37 Cow G.	April 20	+ d sub-max. glands. Cough. Dull and crep. lower r. lung.	"	March 4, 1890. Tuberculosis of lungs and + d glands above udder. Bacilli in both. Udder healthy.
38 Cow G.	Nov. 8		"	" "
39 Cow F.	"	Ibid. Coughs, much.	"	August 21, 1889. Killed. Tuberculosis both lungs, liver, omentum, small intestine. Udder showed one quarter scirrhous. *No bacilli.*
40 Cow I.	"	+ d sub-max. glands. Dull over both lungs. Crepitus on r. side. Cough.	Pos.	March 4, 1890. General condition improved. Lungs almost healthy. Nodules in liver = bacilli. + d glands above udder. Udder slightly scirrhous. *No bacilli.*
41 Cow J.	"	+ sub-max. glands. Both lungs involved. Cough.	"	Jan. 9, 1889. Killed. Tuberculosis of r. lung, liver, and mediastinal gland. *Bacilli in all.* Echinococci in liver.

TABLE I. (*continued*).

Number and Source.	Date.	Symptoms and physical signs of cow.	Result.	Post-mortem, if any made.
42 "Brownie."	Nov. 26		Pos.	By J. F. Winchester. $+$ d lymphatic glands. Lungs generally tuberculous, pleura, mesentery, and ovaries. Udder $=$ scirrhous but not tuberculous.
43 J. F. W.	Nov. 26		Neg.	By J. F. W. Lungs mediastinal and external lymphatics tuberculous. Other parts healthy.
44 Cow D.	Jan. 11	First of milking.	"	Vide supra, 33.
45 Cow D.	Jan. 11	Last of milking.	"	" " "
46 Cow E.	Jan. 16	First of milking.	"	" " 29.
47 Cow E.	Jan. 16	Last of milking.	"	" " "
48 Cow F.	1889 Jan. 18	First of milking.	"	" " 39.
49 Cow F.	"	Last of milking.	"	" " "
50 Cow G.	Jan. 20	First of milking.	"	" " 37.
51 Cow G.	"	Last of milking.	"	" " "
52 Cow H.	Jan. 26	First of milking.	"	" " 31.
53 Cow H.	"	Last of milking.	"	" " "
54 Cow I.	Jan. 30	First of milking.	"	" " 40.
55 Cow I.	"	Last of milking.	"	" " "
56 Cow M.	Feb. 2	First of milking, sediment.	"	
57 Cow M.	"	Last of milking, sediment.	"	
58 Cow L.	March 5	First of milking.	"	June 25, 1890. Large cheesy masses in lungs $=$ Bacilli. Liver mediastinal and ing. glands. Udder healthy.
59 Cow L.	"	Last of milking.	"	" "
60 Cow O.	March 6	First of milking, cream.	Pos.	August 21, 1889. Killed. Lungs, liver. Ing. glands above udder $=$ $+$ d. Udder healthy, except scirrhous in l. post. quarter. No bacilli.

TABLE I. (*continued*).

Number and Source.	Date.	Symptoms and physical signs of cow.	Result.	Post-mortem, if any made.
61 Cow O.	March 6	First of milking, sediment.	Neg.	August 21, 1889. Killed. Lungs, liver. Ing. glands above udder = + d. Udder healthy, except scirrhous in l. post. quarter. No bacilli.
62 Cow O.	"	Last of milking, cream.	"	" "
63 Cow O.	"	Last of milking, sediment.	Pos.	" "
64 Cow P.	"	Cream before death.	"	March 6, 1889. Killed. Whole thoracic cavity and diaphragm, mesentery tuberculous. Liver, ing. gland + d. *Bacilli*. Udder healthy.
65 Cow P.	"	Sediment before death.	Neg.	" "
66 Cow P.	"	Cream after death.	Pos.	" "
67 Cow P.	"	Sediment after death.	"	" "
68 Cow D.	March 11	First of milking, cream.	"	Vide 35.
69 Cow D.	"	First of milking, sediment.	Neg.	"
70 Cow D.	"	Last of milking, cream.	"	"
71 Cow D.	"	Last of milking, sediment.	Pos.	"
72 Cow E.	March 14	First of milking, cream.	Neg.	
73 Cow E.	"	First of milking, sediment.	"	
74 Cow E.	"	Last of milking, cream.	"	
75 Cow E.	"	Last of milking, sediment.	Pos.	
76 Cow F.	March 18	First of milking, cream.	Neg.	
77 Cow F.	"	Last of milking, cream.	"	
78 Cow G.	March 20	First of milking, cream.	"	
79 Cow G.	"	Last of milking, cream.	"	
80 Cow H.	March 23	First of milking, cream.	"	
81 Cow H.	"	Last of milking, cream.	"	
82 Cow L	March 28	First of milking, cream.	Pos.	

TABLE I. (*continued*).

Number and Source.	Date.	Symptoms and physical signs of cow.	Result.	Post-mortem, if any made.
83 Cow I.	March 28	Last of milking, cream.	Neg.	
84 Cow O.	March 30	First of milking, cream.	"	
85 Cow O.	"	Last of milking, cream.	"	
86 Cow M.	April 4	First of milking.	"	June 25, 1890. Nodules all over the skin = Lupus? *Bacilli present.*
87 Cow M.	"	Last of milking.	"	
88 Cow Q.	May 9	First of milking. Cough more or less for a year. Probable general tuberculosis.	"	August 21, 1889. Both lungs and liver. + d ing. gland. Udder healthy.
89 Cow Q.	"	Last of milking. Breathing rapid, poor in flesh, + d = ing. glands, and	"	
90 Slocum.	June 8	F. of milking, sed. Mammaries l. post. quarter of udder.	. "	
91 Slocum.	"	First of milking, cream.	Pos.	
92 Slocum.	"	Last of milking, sediment.	Neg.	
93 Slocum.	"	Last of milking, cream.	Pos.	
94 Saunders.	June 11	First of milking, sed. Cough at times, but general health seems good.	Neg.	
95 Saunders.	"	First of milking, cream.	"	
96 Saunders.	"	Last of milking, sediment.	"	
97 Saunders.	"	Last of milking, cream.	"	
98 Mayhew.	June 18	Emaciated, cough, + d glands in flanks Sediment. Udder healthy.	"	
99 Mayhew.	"	Cream.	Pos.	

TABLE I. (*continued*).

Number and Source.	Date.	Symptoms and physical signs of cow.	Result.	Post-mortem, if any made.
100 Slocum.	June 24	Sediment.	?	
101 Slocum.	"	Cream.	?	
102 Cow R.	July 10	First of milking, sediment.	Neg.	August 21. Killed. Tuberculous deposits in both lungs, spleen, liver. Fœtal membranes and umbilical cord also (about 4 months). Bacilli.
103 Cow R.	"	First of milking, cream.	Pos.	
104 Cow D.	July 13	First of milking, sediment.	Neg.	
105 Cow D.	"	Last of milking, cream.	"	
106 Cow E.	July 19	First of milking, sediment.	"	
107 Cow E.	"	Last of milking, cream.	"	
108 Cow G.	July 23	First of milking, sediment.	"	
109 Cow G.	"	Last of milking, cream.	"	
110 Cow F.	July 25	First of milking, sediment.	"	
111 Cow F.	"	Last of milking, cream.	"	
112 Cow F.	"	First of milking, cream.	"	
113 Cow F.	"	Last of milking, sediment.	"	
114 Cow L.	"	First of milking, sediment.	"	
115 Cow L.	"	First of milking, cream.	"	
116 Cow L.	"	Last of milking, sediment.	"	
117 Cow L.	"	Last of milking, cream.	"	
118 J. F. W.	August 23	Milk.	"	Cow at Lawrence. Udder said to be tuberculous. Attempt at autopsy failed.
119 J. F. W.		Cream.	Pos.	
120	Sept. 5	Milk.	No record.	
121	"	Cream.	"	

A summary of what is shown here is as follows : —
There were 121 examinations of milk and cream made, the
specimens coming from thirty-six different animals. The
bacilli of tuberculosis were found in one or more cover-
glasses upon nineteen different occasions.

These nineteen positive results were obtained from twelve
different animals, and the bacilli were found in about equal
proportion in the milk and the cream ; they were seen more
than once in milk from the same cow, at different examinations, six times.

The bacilli were actually seen, therefore, in specimens from
one third (33%) of the animals examined.

That these animals were actually affected with tuberculosis,
and that the udder was free from disease, was proven in all
possible cases by careful post-mortem examinations. These
were conducted upon twenty out of the thirty-six animals
shown in Table I. and the notes of that examination are
given in the last column of that table.

II.

INOCULATION EXPERIMENTS UPON ANIMALS.

These experiments were conducted under as careful precautions as could be devised. The animals (guinea-pigs and
rabbits) were carefully selected in the first place, and kept
under observation for some time. Any but those apparently
perfectly healthy were rejected, and both before and during
the experiments they were all kept under as perfect hygienic
conditions as could be secured.

The milk and cream used for inoculation was obtained with
the same precautions as was that for the microscopic examinations, was invariably injected subcutaneously, and always,
of course, with a sterilized and fresh syringe for each case.
The animals were kept under observation for at least six
weeks, and were then subjected to exceedingly careful post-
mortem examination.

The results of this work are exhibited in Tables II. (*inoculation experiments upon guinea-pigs*) and III. (*inoculation experiments upon rabbits*).

TABLE II.

INOCULATION OF GUINEA PIGS.

(All Inoculations subcutaneous.)

Number.	Date of Inoculation.	Material used, and Source.	Quantity used.	Date Killed.	Time Elapsed.	Result.
1	1889 Jan. 15	1 of m. Cow D.	8 c. c.	1889 Mch. 1	44 d.	Negative.
2	"	L. of m. Cow D.	7 c. c.	"	"	"
3	"	"	4 c. c.	"	"	"
4	Jan. 19	L. of m. Cow E.	"	"	40 d.	Negative. Pin-head whitish nodules in liver; non-tuberculous.
5	"	1 of m. Cow E.	"	"	"	Negative.
6	"	"	"	"	"	"
7	Jan. 22	1 of m. Cow F.	3 c. c.	Mch. 11	48 d.	Negative. Punctate hemorrhage in lungs; otherwise normal.
8	"	L. of m. Cow F.	"	"	"	" "
9	"	"	1 c. c.	"	"	" "
10	Jan. 26	1 of m. Cow G.	3 c. c.	Jan. 30 Died	4 d.	Negative. Acute peritonitis; plates sterile; cover-glasses showed nothing.
11	"	L. of m. Cow G.	5 c. c.	Jan. 28 Died	2 d.	Negative. Pleurisy and lobar pneumonia. A pig in same pen died of pneumonia on Jan. 24; cultures sterile.
12	"	"	"	Mch. 12	45 d.	Negative. Pregnant.
13	Jan. 29	1 of m. Cow H.	4 c. c.	Mch. 21	51 d.	Negative. Supra-renal capsules apparently enlarged, but negative under the microscope.
14	"	L. of m. Cow H.	10 c. c.	"	"	Negative.
15	"	"	"	"	"	"
16	Feb. 2	L. of m. Cow I.	3 c. c.	Mch. 28	54 d.	"
17	"	1 of m. Cow I.	4 c. c.	"	"	"
18	"	"	4 c. c.	"	"	"
19	Feb. 5	L. of m. Cow M.	2 c. c.	April 3	57 d.	"
20	"	"	"	"	"	"
21	"	"	"	"	"	"

TABLE II. (*continued*).

Number.	Date of Inoculation.	Material used, and Source.	Quantity used.	Date Killed.	Time Elapsed.	Result.
22	Mch. 5.	L. of m. Cow L.	3 c. c.	May 2	58 d.	Negative.
23	"	1 of m. Cow L.	4 c. c.	"	"	"
24	Mch. 9	Cream, Cow P. before death	"	Mch. 18 Died	9 d.	*Positive.* Cheesy mass at point of inoculation; spleen and middle lobe of r. lung congested. *Bacilli* in cover-glasses and sections.
25	"	Cream, Cow P. after death	5 c. c.	May 6	58 d.	*Positive.* Miliary nodules in liver and spleen *containing bacilli.*
26	"	"	"	"	"	*Positive.* Many miliary nodules in liver and spleen *containing bacilli.*
27	"	1 of m. Cream, Cow O.	4 c. c.	Mch. 19 Died.	13 d.	*Positive.* Lungs and mediastinal glands and supra-renal capsule; *bacilli* in all.
28	"	L. of m. Cream, Cow O.	2 c. c.	May 6	58 d.	*Positive.* Miliary nodules in liver and spleen; *bacilli* in both.
29	"	"	"	"	"	Negative.
30	Mch. 18	L. of m. Cream, Cow E.	1 c. c.	May 21	64 d.	Negative. Spleen enlarged. Negative under microscope.
31	"	1 of m. Cow E (sour)	"	"	"	Negative.
32	"	Cow E. (sour)	"	"	"	"
33	"	1 of m. Cream, Cow D.	2 c. c.	"	"	"
34	"	L. of m. Cream, Cow D.	"	"	"	"
35	"	"	1 c. c.	"	"	"
36	Mch. 19	L. of m. Cream, Cow F.	0.6 c. c.	May 31	73 d.	"
37	"	1 of m. Cream, Cow F.	1 c. c.	"	"	*Positive.* Liver and spleen. *Bacilli* in cover - glasses and sections.
38	"	"	"	"	"	Negative.
39	Mch. 23	1 of m. Cow G. (sour)	"	June 5	43 d.	"
40	"	"	"	"	"	"
41	"	"	"	"	"	"

TABLE II. (*continued*).

Number.	Date of Inoculation.	Material used, and Source.	Quantity used.	Date Killed.	Time Elapsed.	Result.
42	Mch. 26	1 of m. Cow H. (sour)	"	June 6	41 d.	Negative.
43	"	"	"	"	"	"
44	"	"	"	April 23 (Died)	28 d.	Negative. *Marasmus.*
45	Mch. 30	L. of m. Cow I. Cream	"	June 6	41 d.	"
46	"	1 of m. Cream, Cow I.	"	"	"	Negative. *Marasmus.*
47	"	"	"	April 8 (Died)	9 d.	Negative.
48	April 2	1 of m. Cream, Cow O.	"	June 6	65 d.	Negative.
49	"	"	"	"	"	"
50	"	"	"	"	"	"
51	"	L. of m. Cream, Cow O.	0.5 c.c.	"	"	"
52	"	"	1 c. c.	"	"	"
53	"	"	"	"	"	*Positive.* Liver. *Bacilli* in cover-glasses and sections.
54	May 11	L. of m. Cream, Cow L.	0.5 c.c.	July 2	52 d.	Negative.
55	"	1 of m. Cream, Cow L. (sour)	1 c. c.	"	"	"
56	"	"	"	"	"	"
57	June 10	Slocum, 1 hr. to u. m. milk.	"	Aug. 1	"	Negative. Many small nodules in spleen; few in liver; not tuberculous.
58	"	"	"	"	"	Negative.
59	"	"	"	"	"	*Positive.* Enlarged glands in flank; nodules in spleen and liver; *bacilli* in gland.
60	June 14	Saunders sed.	"	Aug. 3	50 d.	*Positive.* Spleen enlarged and granular; *bacilli.*
61	"	"	"	"	"	Negative. Spleen enlarged and granular; no bacilli found.
62	"	"	"	"	"	" "
63	June 21	L. of m. Mayhew	"	Aug. 8	48 d.	*Positive.* Spleen enlarged and granular; nodules in liver; *bacilli* in spleen.
64	"	L. of m. Cream, Mayhew	"	"	"	*Positive.* Spleen enlarged and granular; enlarged gland in flank; *bacilli* in gland.

TABLE II. (*continued*).

Number.	Date of Inocula-tion.	Material used, and Source.	Quan-tity used.	Date Killed.	Time Elapsed.	Result.
65	June 21	L. of m. Cream, Mayhew	1 c. c.	Aug. 8	48 d.	*Positive.* Spleen enlarged and granular; enlarged gland in flank; *bacilli* in gland.
66	June 25	Morning, Slocum	"	Nov. 3	131 d.	Negative.
67	"	"	"	"	"	"
68	"	Cream, Slocum	"	"	"	"
69	July 12	1 of m. Cow R.	"	"	114 d.	Negative. Two nodules in ant. surface liver; exami-nation negative.
70	"	"	"	"	"	"
71	"	L. of m. Cream, Cow R.	"	"	"	Negative. Nodules in liver; fibromata.
72	July 16	1 of m. Cow D.	"	Nov. 6	113 d.	Negative.
73	"	L. of m. Cream, Cow D.	"	"	"	"
74	"	"	"	"		Missing.
75	July 19	1 of m. Cow E.	"	"	110 d.	Negative.
76	"	L. of m. Cream, Cow E.	"	"	"	"
77	"	"	"	"	"	"
78	July 25	1 of m. Cow G.	"	Nov. 11	109 d.	"
79	"	L. of m. Cream, Cow G.	"	"	"	"
80	"	"	"	"	"	"
81	July 27	1 of m. Cream, Cow F.	"	Nov. 5	101 d.	"
82	"	L. of m. Cream, Cow F.	"	"	"	"
83	"	"	"	"	"	"
84	July 30	1 of m. Cream, Cow L.	"	"	"	"
85	"	"	"	"	"	"
86	"	1 of m. Beckett	4 c. c.	Nov. 11	104 d.	"
87	Sept. 9.	L. of m. Cream, Beckett	"	"	63 d.	"
88	"	"	5 c. c.	"	"	"

From this table it appears that there were 88 guinea-pigs inoculated with milk from 15 different cows; that tuberculosis was found in 12, and that these results came after the use of milk or cream from six different animals, as follows:

No.	Source.	Material used.	Results found in
24	Cow P.	Milk before death.	Lung.
25	"	Cream after death.	Liver and spleen.
26	"	Cream after death.	L. and s. and renal capsules.
27	Cow O.	Cream.	Glands and lung.
28	"	Cream.	Liver and spleen.
37	Cow F.	Cream.	Liver and spleen.
53	Cow O.	Cream.	Liver.
59	Slocum.	Milk.	Gland.
60	Saunders'.	Milk.	Spleen.
63	Mayhew.	Milk.	Spleen.
64	"	Cream.	Gland.
65	"	Cream.	Gland.

TABLE III.

INOCULATIONS OF RABBITS.

(All inoculations subcutaneous.)

NOTE. — *L. of m.* means that the specimen of milk was taken at the *end* of milking; 1 *of m.* means that it was taken when the milking was begun.

Number.	Date of Inoculation.	Material used and Source.	Quantity used.	Date Killed.	Time Elapsed.	Results.
1	1889 Jan. 15	1 of m. Cow D.	5 c. c.	1889 March 1	44 d.	Negative.
2	"	"	10 c. c.	"	"	"
3	"	L. of m. Cow D.	6 c. c.	"	"	Negative. Small amount fluid in abdomen. Cultures sterile.
4	Jan. 19	L. of m. Cow E.	"	"	40 d.	Negative.
5	"	"	8 c. c.	"	"	"
6	"	1 of m. Cow E.	7 c. c.	Mar. 11	51 d.	Negative. Punctate hemorrhage in lungs.
7	Jan. 22	1 of m. Cow F.	3 c. c.	Mar. 1	38 d.	Negative.
8	"	"	"	Mar. 11	48 d.	"
9	"	L. of m. Cow F.	5 c. c.	"	"	"
10	Jan. 26	1 of m. Cow G.	"	Mar. 12	45 d.	"
11	"	"	"	"	"	"

TABLE III. (*continued*).

Number.	Date of Inoculation.	Material used and Source.	Quantity used.	Date Killed.	Time Elapsed.	Results.
12	Jan. 26	L. of m. Cow G.	10 c. c.	Mar. 12	45 d.	Negative.
13	Jan. 29	1 of m. Cow H.	4 c. c.	Mar. 21	51 d.	"
14	Jan. 29	1 of m. Cow H.	4 c. c.	Mar. 21	51 d.	"
15	"	L. of m. Cow H.	10 c. c.	"	"	"
16	Feb. 2	L. of m. Cow I.	4 c. c.	Mar. 28	54 d.	"
17	"	"	5 c. c.	"	"	"
18	"	1 of m. Cow I.	"	"	"	"
19	Feb. 5	1 of m. Cow M.	2 c. c.	April 3	57 d.	"
20	"	"	3 c. c.	"	"	"
21	"	L. of m. Cow M.	3.5 c. c.	"	"	"
22	Mar. 5	L. of m. Cow L.	5 c. c.	May 2	58 d.	"
23	"	"	6 c. c.	"	"	Negative. Wen under throat.
24	"	1 of m. Cow L.	4 c. c.	"	"	Negative.
25	Mar. 9	Cream before death. Cow P.	5 c. c.	May 6	"	*Positive.* Pin-head nodules in spleen, liver, kidney, and diaphragm. + d gland near liver. *Bacilli* in all.
26	"	"	1 c. c.	"	"	Negative.
27	"	Cream after death. Cow P.	5 c. c.	"	"	*Positive.* Nodule in lungs, liver, spleen, and peritoneum. *Bacilli* in all.
28	"	1 of m. Cream, Cow O.	4 c. c.	"	"	Negative.
29	"	L. of m. Milk, Cow O.	5 c. c.	"	"	"
30	"	L. of m. Cream, Cow O.	4 c. c.	"	"	Negative. Nodule in edge of liver = coccidium oviforme.
31	Mar. 18	L. of m. Cream, Cow E.	1 c. c.	May 28	75 d.	Negative.
32	"	"	"	Died May 27	74 d.	Negative. Marked emaciation.
33	"	Top (sour). Cow E.	"	May 28	72 d.	Negative.
34	"	1 of m. Cream, Cow D.	3 c. c.	"	"	"

TABLE III. (*continued*).

Number.	Date of Inocula-tion.	Material used and Source.	Quan-tity used.	Date Killed.	Time Elapsed.	Results.
35	Mar. 18	Cow D.	5 c.c.	May 28	72 d.	Negative.
36	"	L. of m.	4 c.c.	"	"	"
		Cream, Cow D.				
37	Mar. 19	L. of m.	0.5 c.c.	May 31	73 d.	Negative. Coccidium oviforme in liver.
		Cream, Cow F.				
38	"	"	"	"	"	Negative.
39	"	1 of m.	1 c.c.	"	"	Negative. Bladder worms.
		Cream, Cow F.				
40	Mar. 23	L. of m.	3 c.c.	June 5	74 d.	Negative.
		Cream, Cow G.				
41	"	"	1 c.c.	"	"	Material lost.
42	Mar. 26	1 of m. (sour).	"	"	71 d.	Negative.
		Cow H.				
43	"	"	"	"	"	"
44	"	L. of m. (sour).	"	"	"	Negative. Coccidium oviforme in liver.
		Cow H.				
45	Mar. 30	L. of m.	"	June 6	68 d.	" "
		Cream, Cow I.				
46	May 11	L. of m.	0.5 c.c.	July 2	52 d.	Negative. Fibromata in liver.
		Cream, Cow L.				
47	"	"	.75 c.c.	"	"	*Positive. Liver. Bacilli.*
48	"	1 of m. (sour).	1 c.c.	"	"	Negative.
		Cow L.				
49	June 10	Slocum, 3 hrs. to milking	"	July 31 died	57 d.	Negative. Uræmia. Coccidium oviforme in liver.
50	"	"	"	July 31	"	" "
51	"	"	"	Aug. 1	52 d.	Negative.
52	June 14	Milk, Saunders	"	Aug. 2 died	47 d.	Negative. Apoplexy.
53	"	"	"	"	"	*Positive. Liver. Bacilli.*
54	"	Cream, Saunders	"	Aug. 3	48 d.	*Positive.* Cheesy nodule size of hazel nut at point of inoculation. Glands of abdomen + d. *Bacilli.*
55	June 19	L. of m. Mayhew	"	Aug. 6	"	Negative. Coccidium oviforme in liver.
56	"	"	"	Aug. 3 died	45 d.	" "
57	June 21	L. of m. Cream, Mayhew	"	Aug. 8	48 d.	Negative.

TABLE III. (continued).

Number.	Date of Inoculation.	Material used and Source.	Quantity used.	Date Killed.	Time Elapsed.	Results.
58	June 25	Milk, Slocum	1 c. c.	Nov. 3	131 d.	Negative.
59	"	"	"	"	"	"
60	"	Cream, Slocum	"	"	"	"
61	July 12	1 of m. Cow R.	1 c. c.	"	114 d.	"
62	"	"	"	Aug. 3	22 d. Died.	Decomposition too rapid for examination.
63	"	L. of m. Cream, Cow R.	"	Aug. 5	24 d.	Negative. Death from constipation.
64	July 16	1 of m. Cow D.	"	Nov. 6	113 d.	Negative. Coccidia in liver.
65	"	"	"	· "	"	Negative.
66	"	L. of m. Cream, Cow D.	"	"	"	Negative. Perihepatitis.
67	July 19	1 of m. Cow E.	"	"	110 d.	Negative.
68	"	"	"	"	"	"
69	"	L. of m. Cow E.	"	"	"	"
70	July 23	1 of m. Cow G.	"	Aug. 16	22 d.	Negative. Rupture of bladder. Specimens lost.
71	"	"	"	Nov. 11	109 d.	Negative.
72	"	L. of m. Cow G.	"	"	"	"
73	July 27	1 of m. Cream, Cow F.	"	Nov. 5	101 d.	"
74	"	"	"	"	"	"
75	"	L. of m. Cream, Cow F.	"	"	"	"
76	July 30	1 of m. Cream, Cow L.	"	Aug. 13	14 d.	Negative. Killed by owl and buried at farm without autopsy.
77	"	"	"	Nov. 5	48 d.	Negative.
78	Sept. 9	1 of m. Beckett	5 c. c.	Nov. 11	64 d.	"
79	"	"	4 c. c.	"	"	"
80	"	L. of m. Beckett	8 c. c.	"	"	Positive. Nodule size of a pea at point of inoculation. Bacilli in sections and cover-glasses.
81	1890 June 18	Cream, Cow W.	3 c. c.	Aug. 6	49 d.	Negative.
82	"	"	2 c. c.	"	· "	"
83	"	"	4 c. c.	"	"	"
84	"	Cream, Pierce, ant. teat	2 c. c.	"	"	"

TABLE III. (*continued*).

Number.	Date of Inocula- tion.	Material used and Source.	Quan- tity used.	Date Killed.	Time Elapsed.	Results.
85	June 18	ant. teat	4 c. c.	Aug. 6	49 d.	Negative.
86	"	"	"	"	"	"
87	"	Ibid., hind teat	5 c. c.	"	"	"
88	"	Ibid., hind teat	4 c. c.	"	"	"
89	"	"	5 c. c.	"	"	"
90	"	Cream, Cow Y.	3 c. c.	"	"	"
91	"	"	2.5 c. c.	"	"	"
92	"	"	"	"	"	"
93	"	Cream, Cow X.	5 c. c.	"	"	"
94	"	"	"	"	"	Negative. Coccidia in liver.
95	"	"	6 c. c.	"	"	Negative. Cheesy nodule at point of inoculation and + d gland near. No ba- cilli found.

From this table it appears that 95 rabbits were used for the same purposes and under the same conditions as were the guinea-pigs in Table II. Of these rabbits five (Nos. 41, 62, 63, 70, and 76) were for various reasons useless for the purposes of the investigation, leaving 90 which were subjected to full examination. For these 90 animals milk from 19 different cows was used one or more times, and tuberculosis was found in 6 animals inoculated with milk from four different cows, as follows: —

No.	Source.	Material used.	Results found in
25	Cow P.	Cream before death.	Spleen, kidney, liver, diaphragm.
27	"	Cream after death.	Spleen, kidney, liver, diaphragm.
47	Cow L.	Cream.	Liver.
53	Saunders'.	Milk.	Gland.
54	"	Cream.	Gland.
80	Beckett.	Cream.	Cæcum.

These results show a less proportion of apparent infection of the milk as demonstrated by the inoculation experiments than appeared to be the case in the microscopic examinations. But this difference, even granting that they were *all* the results of the inoculations, is no more than might be expected and explained by causes beyond control.

III.

The third line of experiment was in feeding the milk from tuberculous cows and healthy udders to different series of animals. Here again the greatest precautions were taken against outside infection, and it is believed that these were as free from sources of error as it is ever possible to make such experiments. They were carried on upon rabbits, pigs, and calves, and the statement of the experiments is shown in tables IV., V., and VI.

TABLE IV.

MILK-FEEDING EXPERIMENTS UPON RABBITS.

Number.	Date.	Fed with Milk from	Killed.	Result.
1	Dec. 3, 1889	Cow D, m. & e.	May 4, 1890	None.
2	"	"	"	"
3	"	"	"	"
4	"	"	May 20	"
5	Aug. 28, 1888	Cow I, m. & e.	Feb. 16, 1889	"
6	"	"	"	"
7	"	"	"	"
8	"	"	"	"
9	"	"	"	"
10	"	"	"	"
11	"	"	"	Nodules in liver. Material lost.
12	Feb. 16, 1889	Cow E, m. & e.	Mar. 19 Died	Died while pregnant. Both lungs full of miliary tubercles. Showing *bacilli*.
13	"	"	June 8	Coccidia in liver.
14	"	"	"	" "
15	"	"	"	" "
16	Feb. 25, 1889	Cow O, irregularly.	Sept. 14 Died	Acute pneumonia.
17	"	"	Sept. 30	None.
18	"	"	"	"
19	"	"	"	"
20	"	"	"	Coccidia in liver.
21	"	"	"	" "
22	"	"	"	" "

TABLE IV. (*continued*).

Number.	Date.	Fed with Milk from	Killed.	Result.
23	Dec. 17, 1889	Cow E, m. & e.	Feb. 16, 1890 Died	Few nodules in liver = small inflammatory nodule.
24	"	"	Mar. 23	None.
25	"	"	Mar. 24	Hæmatoma of liver.
26	"	"	Mar. 25	Few spots in lung. Active hyperæmia.
27	"	"	Mar. 24	Congestion of lung and miliary tubercle, — no bacilli of tuberculosis.
28	"	"	"	Miliary nodule in liver showing B. T.
29	"	"	"	None.
30	"	"	Mar. 31	Material thrown away by accident.
31	"	"	April 7	Miliary nodules of liver = micrococci.
32	"	"	May 14	None.
33	"	"	"	"
34	"	"	"	"
35	"	"	"	"
36	"	"	"	"
37	"	"	"	"
38	"	"	"	"
39	"	"	"	"
40	Feb. 5, 1890	Cow V, m. & e.	May 15	Coccidia in liver.
41	"	"	"	None.
42	"	"	"	"
43	"	"	"	"
44	"	"	"	"
45	"	"	"	"
46	"	"	"	"
47	"	"	"	"
48	"	"	"	"

Forty-eight animals experimented upon. Two showed positive results. Both fed upon milk of Cow E.

There were used 48 animals, with positive results (tuberculosis) in two, and both of these animals were fed upon milk from cow E, No. 12, one nodule in lung, after 31 days; No. 28, one nodule in liver after 97 days.

This is of course a very small proportion of positive results, but the following table shows a very different condition of affairs, that is especially striking for the reason that pigs are not believed to be unusually susceptible to tuberculosis under ordinary conditions.

TABLE V.

MILK-FEEDING EXPERIMENTS UPON PIGS.

No.	Age.	Date.	Fed on Milk from.	Killed.	Result of Post-mortem.
1	8 weeks	Mar. 28, '88	Cow E for one week, then cow F and Surplus.	July 26, '88	Nodules in the liver in which *tubercle bacilli* were found, and a pleuritic adhesion on left side.
2	"	"	"	"	Nodule in left lung, nodules in liver, and enlarged submaxillary lymphatic, *tubercle bacilli* found in all.
3	"	"	"	Sept. 26, '88	Nodules in liver, which, however, were not saved.
4	"	"	"	"	Nodules in liver, which, however, were not saved.
5	"	"	"	Dec. 11, '88	Negative.
6	"	"	"	"	"
7	"	"	"	"	Two little nodules in spleen, in which *tubercle bacilli* were found.
8	"	Mar. '89	Surplus milk	Nov. 21, '89	A small nodule in the liver, in which *tubercle bacilli* were found.
9	"	"	"	"	A few yellow spots in liver, found to be enlarged blood vessels.
10	"	"	"	"	A small nodule in the liver, in which *tubercle bacilli* were found. *Plate.*
11	10 weeks	Dec. 12, '89	Cows E & L	May 3, '90	Enlarged mesenteric glands, and nodules in the liver; the latter were granulation tissue, but no tubercle bacilli were found in either.
12	"	"	Cows M & W	"	Enlarged submaxillary lymphatic gland, and nodules in the liver, latter composed of granulation tissue. No tubercle bacilli found in liver, gland?

Twelve healthy animals were used with positive results (demonstration of the bacilli under the microscope) in five. In two others, nodules presenting the gross appearance of tuberculosis were found, but the material was not saved for microscopic examination. In any case, nearly fifty per cent. of the animals were shown to be tuberculous, as follows : —

No.	Material used.	Results found in
1	Cows E and F.	Liver.
2	Cows E and F.	Liver, lung, and gland.
7	Cows E and F.	Spleen.
8	Surplus Milk.	Liver.
10	Surplus Milk.	Liver.

For the purposes of the third series of feeding experiments, calves were bought as young as possible, and from as healthy parentage as could be found.

There were twenty-five calves used in this series of experiments, but of these four (G, T, U, and X) are to be excluded from the count, leaving 21.

TABLE VI.

MILK-FEEDING EXPERIMENTS UPON CALVES.

No.	Age and Date.	Parentage and Source.	Fed on Milk from.	Killed.	Result of Post-mortem.
A.	6 days old Feb. 18, '88	Grade Holstein, healthy. Mattapan. Heifer.	Cow A to Feb. 27. Then Cows A & B to Mar. 19. Then Cows A & C until killed.	Sept. 26,'88	Nodules found in right lung, liver and an enlarged mediastinal lymphatic. *Tubercle bacilli* found in lung. Other organs ?
B.	4 days old Feb. 29,'88	Red native heifer, healthy. Mattapan.	Cows B & C Feb. 29 to Apr. 8, then Cows B & H.	July 6, '88.	Enlarged mediastinal and pharyngeal lymphatics, nodules in liver and two nodules in anterior lobe of right lung. *Tubercle bacilli* found in sections of lung. See plate.

TABLE VI. (*continued*).

No.	Age and Date.	Parentage and Source.	Fed on Milk from.	Killed.	Result of Post-mortem.
C.	3 days old Apr. 3, '88	Healthy. Jamaica Plain.	Cow E.	Sept. 26,'88	Slightly enlarged mediastinal lymphatic. Not tuberculous.
D.	3 days old Apr. 9, '88	Healthy. Jamaica Plain.	Cow D.	"	A few nodules in the liver, in which *tubercle bacilli* were found. *See plate.*
E.	3 days old July 9, '88	Healthy. Roxbury.	Cow J.	Jan. 9, '89	None.
F.	2 days old Sep. 24, '88	Healthy. Jamaica Plain.	Cow I.	Apr. 10, '89	A few nodules in liver and kidney, *tubercle bacilli* found in latter.
G.	5 days old Oct. 19, '88	Healthy. Mattapan.	Cows D & G.	"	A few small nodules in liver, specimen lost.
H.	10 days old Dec. 4, '88	Healthy. Canterbury.	Cows E & F.	"	A nodule at lower border of liver, in which *tubercle bacilli* were found.
I.	1 day old Feb. 18,'89	Healthy. Mattapan.	Cow L.	Aug. 13,'89	Two small white spots in liver, and a mottled appearance of one kidney. Kidney negative. Liver (?)
J.	2 days old Feb. 25,'89	Healthy. Jamaica Plain.	Cow O, helped out by healthy Cow K.	"	Enlarged mesenteric lymphatics in which *tubercle bacilli* were found. Kidney like calf I's, negative.
K.	1 day old Mar. 29,'89	Healthy (?) mother. Mattapan.	Cow M from Apr. 13.	"	Chronic interstitial pneumonia right lung, nodule in liver, congested kidney. Not tuberculous. *Plate.*
L.	5 days old May 1, '89	Healthy. Brookline.	Cow Q (calf had cough June 3 to Aug. 1).	"	Negative.
M.	2 days old May 3, '89	Healthy. Jamaica Plain.	Cow R (calf had cough June 18 to Aug. 1).	"	Nodules in liver in which *tubercle bacilli* were found. Kidneys, negative.
N.	7 days old May 16,'89	Healthy. Jamaica Plain.	Cow D.	Nov. 21, '89	Nodules in liver, and slightly enlarged mesenteric glands. No tubercle bacilli found in either.
O.	6 days old May 23,'89	Healthy. Jamaica Plain.	Cow F to June 1, then Cow E.	"	One nodule in liver, negative. Mesenteric lymphatic (?)
P.	5 days old July 18,'89	Healthy. Mattapan.	Cow S.	Mar. 4, '90	Negative.

TABLE VI. (*continued*).

No.	Age and Date.	Parentage and Source.	Fed on Milk from.	Killed.	Result of Post-mortem.
Q.	3 days old Aug. 3, '89	Healthy. Jamaica Plain.	Cow L.	Mar. 4, '90	A few nodules in liver, negative. A hair ball about size of a base ball in the rumen.
R.	11 days old Aug 22,'89	Healthy. Mattapan.	Cow G.	"	A few nodules in the liver. Not tuberculous.
S.	4 days old Oct. 22, '89	Healthy. Jamaica Plain.	Cow V.	"	Nodules in liver, negative. Slightly enlarged pharyngeal lymphatics?
T.	5 days old Mar. 11,'90	Healthy. Mattapan.	Cow X.	June 25, '90	Negative. Cow X proved to be not tuberculous.
U.	5 days old Mar. 11,'90	"	Cow L.	"	Enlarged mesenteric glands.
V.	3 days old Mar. 18,'90	"	Cow M.	"	Enlarged spleen given to Dr. Jeffries. Contained no tubercle bacilli.
W.	14 days old Apr. 7, '90	"	Cow S.	"	Cheesy nodule in lung, in which *tubercle bacilli* were found. Red spot in liver?
X.	4 days old Apr. 7, '90	"	Cow Y.	Apr. 16 Moribund & killed.	Pneumonia and pleurisy, enlarged mediastinal lymphatics, enlarged spleen. Dr. Jeffries found his swine disease organism in these specimens.
Y.	3 days old May 31,'00	"	Cow Y.	June 25, '90	Small yellow spot in liver. Microscopic ex. = negative.

Of these twenty-one animals, eight, or over 33%, were shown to be tuberculous, as follows : —

Letter.	Milk.	Result in
A	Cow A–A and B, F.	Lung.
B	Cow B and C, B and H.	Lung.
D	Cow D.	Liver.
F	Cow F.	Kidney.
H	Cow E and F.	Liver.
J	Cow O.	Gland.
M	Cow R.	Liver.
W	Cow S.	Lung.

It is of course true that pigs and calves, that drink milk much more freely than do rabbits, are more susceptible to infection by the gastro-intestinal tract, and that this may explain the far greater proportion of positive results in these two species of animals.

That the cows from which the milk for these feeding experiments was derived were free from tuberculosis of the udder, is shown by the following table of their histories, and the results of the post-mortem examinations.

TABLE VII.

Cows used for Experiment.

No.	Date.	Kind.	Source.	Killed.	Condition at Purchase.	Result of Post-mortem.
A.	Feb. 16, 1888	Red and white. Native.	Tuberculous herd. Peabody Poor Farm.	Jan. 9, 1889	Enlarged sub-maxillary lymphatics; nodulated udder.	Right lung tuberculous; tubercle bacilli found in cover-glass preparations; enlarged inguinal lymphatic; udder, thickened walls of ducts and sinuses, but no tuberculosis.
B.	Feb. 27, 1888	High grade Jersey.	J. S. Russell, Milton.	July 6, 1888	Cough; expectoration; pulmonary tuberculosis; udder healthy.	General tuberculosis of lungs, pleura, liver, peritoneum, capsule of spleen, both ovaries, uterus, and lymphatic glands; udder scirrhous, but no local deposits; tubercle bacilli found in lungs.
C.	Feb. 28, 1888	10 or 12 years old. Red native.	P. McMorrow, Jamaica Plain.		Udder nodulated, but not sure of tuberculosis.	Not used for tuberculosis experiments.
D.	March 22, 1888	11 years old. Black native.	Danvers Insane Asylum.	Nov. 21, 1889	Enlarged sub-maxillary glands; friction sound over right lung; no dullness; April 9, udder scirrhous.	Enlarged inguinal lymphatic; no tubercle bacilli found in it; scirrhous udder; no tubercle bacilli found; cheesy nodule in liver, and abscesses in both lungs; in both liver and lung tubercle bacilli were found.
E.	"	12 years old. Red native.	"	June 21, 1890	Cough; enlarged sub-maxillary lymphatic glands; crepitus and dullness on lower part of right lung; Feb. 15, 1889, right hind quarter of udder indurated.	Anterior lobe of right lung a tuberculous mass; abscesses in upper part of posterior lobes of both lungs, in which tubercle bacilli were found; udder in right hind quarter contained a few nodules, in which only cocci were found; tubercle bacilli found in a small abscess opposite an opening made in trachea the previous spring.

F.	March 22, 1888	11 years old. Grade short-horn.	Danvers Insane Asylum.	Aug. 25, 1889	Cough; enlarged sub-maxillary glands; râles and dullness low down on right side.	Tubercles in both lungs, in which bacilli of tuberculosis were present; nodules in liver, mesentery and small intestines, in which tubercle bacilli were found; minute nodules in udder, in which one tubercle bacillus in a *giant cell* was found.
G.	"	9 years old. Brindle and white native.	"	March 4, 1890	Cough; enlarged sub-maxillary lymphatics; râles and dullness lower part of right lung.	Tuberculosis of lungs, and enlarged lymphatics above udder; tubercle bacilli were found in both; *udder healthy.*
H.	"	15 years. Red native.	"	April 10, 1889	Enlarged sub-maxillary lymphatics; both lungs probably tuberculous; left hind quarter of udder slightly indurated.	Few tubercles in left lung; enlarged mesenteric lymphatics; small nodules in intestines; cover-glass preparations showed tubercle bacilli; *udder healthy.*
I.	"	9 years old. Red native.	"	March 4, 1890	Enlarged sub-maxillary glands; dullness low down on both lungs; râles over right lung; cough; June 10, 1889, some slight induration of right hind quarter of udder.	Condition had improved much over first arrival; lungs almost healthy; two or three nodules in liver; some enlarged lymphatics above udder; udder showed a slight increase in fibrous tissue; tubercle bacilli found in nodules from liver.
J.	April 5, 1888	6 years old. Red native.	Peabody, Mr. J. C. Rogers.	Jan. 9, 1889	Enlarged sub-maxillary glands; cough; probably both lungs affected.	Tuberculosis in right lung; nodules in liver; tubercle bacilli found in both; *udder healthy.*
K. L.	March 31, 1888	Red native.	Jamaica Plain, Mr. Motley.	June 25, 1890	Cough.	Not used for tuberculosis experiments. Large cheesy masses in posterior lobes of both lungs, in which tubercle-bacilli were found; many nodules in liver; enlarged mediastinal gland; enlarged glands above udder; *udder healthy.*
M.	Dec. 22, 1888	5 or 6 years grade, Guernsey.	Peabody, Mr. J. C. Rogers.	"	Breathes badly; nodules all over skin; lupus (?).	Enlarged cheesy lymphatic gland back of pharynx; no tubercle bacilli found in it; June 10, 1889, tubercle bacilli found in some of nodules under skin.

TABLE VII. (*continued*).

No.	Date.	Kind.	Source.	Killed.	Condition at Purchase.	Result of Post-mortem.
N.	Dec. 20, 1888	Guernsey.	Peabody, Mr. J. C. Rogers.	Jan. 9, 1889		A few nodules in lungs and liver, and an enlarged mediastinal gland, in which tubercle bacilli were found.
O.	Feb. 21, 1889	8 or 9 years old. Jersey.	Newport, R. I. C. Vanderbilt.	Aug. 21, 1889	General tuberculosis.	Nodules in lungs, liver, posterior mediastinal glands; enlarged glands above udder; tubercle bacilli found in all; udder healthy, except a slight increase in its fibrous tissue.
P.	March 1, 1889	10 or 12 years old. Grade Jersey.	Framingham, O. S. Robinson.	March 6, 1889	Old general tuberculosis; udder healthy; very weak; milk used as a supply.	General advanced tuberculosis all through thoracic cavity; slight in liver and inguinal glands near udder; tubercle bacilli found in all; milk contained tubercle bacilli; *udder healthy.*
Q.	April 22, 1889	5 or 6 years old. Pure Guernsey.	Brookline, Mr. Knapp.	Aug. 21, 1889	General tuberculosis; dullness over right lung	Tuberculosis of both lungs; few nodules in liver; enlarged inguinal gland above udder; all show tubercle bacilli; *udder healthy.*
R.	"	4 or 5 years old. Jersey.	"	"	Phthisis; dull over left lung.	Tuberculosis in posterior part of both lungs; few nodules in capsule of spleen, in liver; enlarged inguinal gland above udder; udder scirrhous; nodules on membranes and umbilical cord; fœtus healthy; tubercle bacilli found in lung, liver, spleen, and gland; *udder healthy.*
S.	June 25, 1889	High grade Guernsey.	Jamaica Plain, W. H. Slocum.	June 26, 1890	Bad breathing; enlarged flank glands; slight cough; mammitis, which had disappeared by Aug. 8.	Small nodules in liver, and enlarged glands above udder; tubercle bacilli found in both; *udder healthy.*

T.	August 22, 1889	7 or 8 years old. Jersey.	Wellesley, S. M. Weld.	Nov. 5, 1889	Cough; wheezing respiratory murmur over left lung; enlarged inguinal and sub-maxillary glands; udder indurated.	Tuberculosis both lungs, ribs, mediastinal and thoracic glands, and glands above udder and elsewhere; nodules in liver; tubercle bacilli found in them; udder showed only an increase in connective tissue; contained no B. T.
U.	Sept. 2, 1889	3 years old. Grade Guernsey.	Barre, W. H. Bowker.	Oct. 3, 1889	Enlarged cervical lymphatic glands and enlarged glands at flank; udder healthy.	Tuberculosis of all glands and slight tuberculosis of lungs and liver; tubercle bacilli found in them; udder scirrhous (?).
V.	Oct. 19, 1889	Native.	Cambridge, Dr. J. L. Hildreth.	June 21, 1890	Cough; enlarged sub-maxillary glands, and at flank.	Tuberculous pleuritis; nodules in liver; tubercle bacilli found; abscesses at base of two teats contained cocci, but no tubercle bacilli.
W.	Nov. 2, 1889	8 years old. Jersey.	Lynnfield, H. Saltonstall.	"	Cough for 4 months.	Enlarged inguinal glands; nodule in right lung and liver; tubercle bacilli found in lung, but not in gland.
X.	March 11, 1890	11 or 12 years old. Jersey.	Jamaica Plain, W. H. Slocum.	June 26, 1890	Cough for a year or two, and slightly enlarged glands at throat and flank.	Emphysema of lungs; enlarged glands above udder; presented no evidence of tuberculosis.
Y.	April 4, 1890	Ayrshire.	Brookline, C. S. Sargent.	"	Cough; moist râles upper part of left lung.	Lungs healthy; nodules in liver, in which tubercle bacilli were found; udder healthy.

Examination of Cows used for Experiment.

Twenty-three of the twenty-five cows shown upon this table were used for the feeding experiments, and in not one of them did the most careful macroscopic and microscopic search show any sign of tuberculosis of the udder except in one, Cow F, and in this case a single giant cell, containing one bacillus, was found in one section, and no other indication of tuberculosis anywhere else in the udder. In all of these cows, however, tuberculosis was demonstrated to be present in some other part of the body than the udder.

An exceedingly interesting piece of evidence as regards the hereditary nature of tuberculosis is found in the history of nineteen calves born of these tuberculous cows with healthy udders, and shown in the following table.

TABLE VIII.

CONDITION OF CALVES FROM TUBERCULOUS COWS KEPT AT FARM.

No.	Mother.	Born.	Killed.	Result of Post-mortem.
1	Cow D	1888 Apr. 5	1888 Apr. 6	Perfectly healthy.
2	D	1889 Apr. 29	1889 May 4	" "
3	E	1888 March 23	1888 March 24	Fœtal membranes covered with nodules. Slight atalectasis of left lung. Mesenteric lymphatics large, but no more so than usual in a young animal.
4	E	1889 May 26	1889 June 1	Healthy.
5	E	1890 June 18	1890 June 21	"
6	F	1888 May 14	1888 May 14	Healthy, enlarged mesenteric glands, but normal.
7	F	1889 May 5	1889 May 9	Healthy.
8	G	1888 Apr. 15	1888 Apr. 16	Healthy, except a few nodules on pericardium and peritoneum. (Did not look tuberculous. Material lost.) (Hæmolymph glands?)
9	G	1889 July 15	1889 Nov. 21	Healthy, except a few nodules on edge of liver, found to be nothing abnormal.
10	H		1889 Aug. 10	A five months' fœtus. Nodules on membranes and cord, fœtus healthy. No record of micros. ex.

TABLE VIII. (continued).

No.	Mother.	Born.	Killed.	Result of Post-mortem.
11	I	1888 July 11	1888 July 12	Healthy, except mesenteric lymphatics appeared large. (Not more so than many others that are normal.)
12	I	1889 July 25	1889 July 31	Healthy.
13	I		1890 March 4	A six months' fœtus, normal except nodules on membranes.
14	J	1888 June 15	1888 June 16	Healthy, mesenteric lymphatics appeared large.
15	L	1890 March 3	1890 March 18	Healthy.
16	M	1889 March 30	1889 Apr. 4	"
17	M	1890 March 15	1890 March 18	"
18	O	1889 May 31	1889 June 4	Calf died, death due to bronchitis, healthy as far as tuberculosis is concerned.
19	S	1890 March 31	1890 March 31	Healthy.

Of these nineteen calves, all killed within six days after birth, not one showed any detectable evidence of tuberculosis, and a most careful search was made in all cases. So that this certainly seems to point away from any very active transmission of tuberculosis from the cow to its offspring.

IV.

As an interesting corollary to the work already detailed, a series of microscopic examinations and inoculation experiments were made with milk taken at random from the mixed supply of the city of Boston. The samples were obtained from the Inspector of Milk, and the work done is exhibited in Tables IX. and X.

TABLE IX.

COVER-GLASS EXAMINATION OF MILK AND CREAM FROM MILK SUPPLY OF CITY OF BOSTON.

No.	Source.	Date.	Result.
1	Milk 1. Cream.	Feb. 15	Negative.
2	Milk 1. Sediment.	"	"
3	Milk 2. Cream.	"	"
4	Milk 2. Sediment.	"	"
5	Milk 3. Cream.	"	"
6	Milk 3. Sediment.	"	"
7	Milk 4. Cream.	"	"
8	Milk 4. Sediment.	"	"
9	Milk 140. Cream.	Feb. 22	"
10	Milk 30. Cream.	"	"
11	Milk 270. Cream.	"	"
12	"Draper." Cream.	"	"
13	Milk 468. Cream.	"	"
14	Milk 391. Cream.	"	"
15	5683 d. Milk.	"	"
16	3697 d. Milk.	March 1	"
17	3687 d. Milk.	"	"
18	3701 d. Milk.	"	"
19	"Lowell." Cream.	March 6	"
20	"Lowell." Milk.	"	"
21	3849 d. Cream.	March 10	"
22	3849 d. Milk.	"	"
23	3845 d. Cream.	"	"
24	384 d. Milk.	"	"
25	3851 d. Cream.	"	"
26	3851 d. Milk.	"	"
27	3847 d. Cream.	"	"

TABLE IX. (*continued*).

No.	Source.	Date.	Result.
28	3847 d. Milk.	March 10	Negative.
29	3973 d. Cream.	March 14	"
30	3973 d. Milk.	"	"
31	3979 d. Cream.	"	"
32	3979 d. Milk.	"	"
33	4125 d. Cream.	March 26	"
34	4125 d. Milk.	"	"
35	4121 d. Cream.	"	"
36	4121 d. Milk.	"	"
37	4123 d. Cream.	"	"
38	4123 d. Milk.	"	"
39	4315 d. Cream.	April 3	"
40	4315 d. Milk.	"	"
41	4325 d. Cream.	"	Bacilli found.
42	4325 d. Milk.	"	Negative.
43	4637 Cream.	April 16	"
44	4637 Milk.	"	"
45	4617 Cream.	"	"
46	4617 Milk.	"	"
47	4629 d. Cream.	"	"
48	4629 d. Milk.	"	"
49	4619 d. Cream.	"	"
50	4619 d. Milk.	"	"
51	4797 d. Cream.	April 28	"
52	4797 d. Milk.	"	"
53	4809 d. Cream.	"	"
54	4809 d. Milk.	"	"
55	4815 d. Cream.	"	"
56	4815 d. Milk.	"	"

The table shows that there were fifty-six examinations made of the milk and cream from thirty-three samples, with the result of demonstrating the presence of the bacilli of tuberculosis once. (No. 41.)

TABLE X.

INOCULATIONS WITH MILK FROM MILK SUPPLY OF THE CITY OF BOSTON.
RABBITS.

No.	Date Inoc-ulated.	Source of Material.	Quantity Used.	Date Killed.	Result.
1	1890 Feb. 22	Cream 140, Subcu.	3 c. c.	May 8	Negative.
2	"	"	"	"	"
3	"	Cream 30,	"	"	Bladder worms; nodule in liver; granulation tissue and bacilli.
4	"	"	"	"	Negative.
5	"	Cream 270,	"	May 12	Negative.
6	"	"	2 c. c.	"	Enlarged mesenteric gland and nodule in liver; no bacilli; coccidia in liver.
7	"	Cream from "Draper's" milk.	1 c. c.	"	Negative; great emaciation.
8	"	"	"	Died Feb. 25	No autopsy.
9	March 1	Sediment of 5683 d.	4 c. c.	May 12	Bladder worms; nodules in liver, and two in spleen; nodules cheesy; but no bacilli, and no coccidia.
10	"	"	"	"	Nodules in cæcum; cover-glasses and sections showed coccidia and *bacilli of tuber-culosis.*
11	"	Sediment of 3697 d.	3 c. c.	Died April 8	Negative.
12	"	"	"	Died March 5	"
13	"	Sediment of 3687 d.	4 c. c.	May 12	"
14	"	"	3 c. c.	"	Negative. Coccidia in cæcum.
15	"	Sediment of 3701 d.	4 c. c.	May 15	Negative.
16	"	"	"	"	Negative. Bladder worms.
17	March 6	Cream, "Lowell" milk.	4 c. c.	May 8	Negative.
18	"	"	"	"	Enlarged spleen, and nodules in cæcum and liver; coccidia in last two.

TABLE X. (*continued*).

No.	Date Inoc- ulated.	Source of Material.	Quantity Used.	Date Killed.	Result.
19	March 6	Sediment of " Lowell " milk.	4 c. c.	May 8	Negative.
20	"	"	"	Died May 5	Acute peritonitis from rupture; spleen much enlarged, and many nodules all over intestinal wall; bacilli in spleen; *plate photograph.*
21	Mch. 11	Cream 3845 d.	3½ c. c.	May 15	Nodules in cæcum; coccidia.
22	"	"	"	"	Negative.
23	"	Cream 3849 d.	4 c. c.	"	Negative. Calcareous nodule in liver not examined.
24	"	"	"	"	Negative. Nodules in cæcum; negative.
25	Mch. 12	Upper B. Robinson.	4 c. c.	"	Negative. Few nodules in liver and cæcum; *material lost.*
26	"	"	"	"	Negative. Cæcum coccidia; some infiltration at point of inoculation.
27	"	"	"	Died March 12	Acute general peritonitis; negative.
28	"	"	"	May 15	Yellow nodule at point of inoculation; *no b.* Cæcum coccidia; liver, small nodule; no bacilli; emaciation.

In this table the result of the inoculations of this milk is shown, and by it it appears that there were twenty-eight rabbits used, of which three (Nos. 8, 25, and 27) are to be excluded, leaving twenty-five in which the investigation was completed. Among these twenty-five there were positive results in three, as follows: —

Number.	Material.	Results.
3	Cream 30.	Liver.
10	Milk, 5683 d.	Cæcum.
20	Lowell milk.	Spleen.

Of course these results, obtained in milk from a mixed source, are not as conclusive upon the especial point toward

which the main line of investigation was directed, — as to
the presence of the bacilli of tuberculosis in milk from cows
with healthy udders, — but they certainly tend to demonstrate that there may be this infectious element in any milk
supply from uninspected cattle.

V.

The last of the lines in which investigation was made was
to endeavor, if possible, to obtain clinical reports of cases of
transmission through milk from mother to offspring, and evidence was sought in this direction as follows : —

In January and February of the year 1890 a circular was
sent out to about eighteen hundred medical and veterinary
gentlemen, in an attempt to discover any clinical cases bearing upon the subject at hand. The list was chosen, in the
first place, from the members of the Massachusetts Medical
Society of at least five years' standing, and was then filled
out with the names of the members of the American Surgical
Association, the Association of American Physicians, and one
or two of the other special societies of the country. The
names of the veterinarians were taken from the rolls of the
United States Veterinary Association, and included those
who were thought to have had enough experience to make
their observation of possible value in this direction, in the
same way as the list of medical men was completed. A copy
of the circular follows : —

HARVARD MEDICAL SCHOOL, BACTERIOLOGICAL LABORATORY,
BOSTON, January, 1890.

DEAR SIR, — It is desired to obtain a collection of statistics upon the following point: Have you ever seen a case of
Tuberculosis which it seemed possible to you to trace to a
milk supply as a cause ?

An answer upon the inclosed postal card will greatly oblige

Yours very truly,

HAROLD C. ERNST, M. D.

Dr. ———.

With each one of these circulars was inclosed a postal card with my address printed upon it, so that the way for an answer should be made as easy as possible.

The object in sending out this circular letter was not in the hope of obtaining many exact observations, for it must be granted at once, and without argument, that the clinical reports of such cases as are here inquired after must be of small value from the point of view of experimental science; it was, however, our wish to see if there were an opinion among the medical profession at large in favor of such a source of tuberculosis, and if so, how far that opinion extended. The results obtained seem to have justified the time and expense of the investigation; for, of all the replies received, but an extremely small number have expressed a disbelief in the possibility of such an origin of the disease, a very large number have shown how widespread the suspicion of it has extended, and a considerable number have replied that they have either suspected such an origin or give cases to exemplify it. This appears to be the more remarkable, for the reason that even the infectious nature of tuberculosis has been so little suspected in some parts of the country until recently, and still more so because the discovery of the infectious agent and the scientific proof of its power has been a matter of so short a time.

In all cases in which there seemed to be a loophole, from the form of the answer on the postal card, to think that the writer had suspected the existence of such a case in his own or a friend's practice, a letter was sent asking for further details; to most of these, however, there was either no reply, or else it was said that nothing was meant by the form of expression used. To those gentlemen who took the trouble to answer the inquiries sent them our thanks are certainly due, and are rendered with pleasure.

The correspondence is given somewhat fully, but purposely so, in order to show, as completely as may be, the opinion of the medical profession at large upon this question. Every one of the objectors is quoted in full, although none of them gave any very full reasons for their disbelief in this method of transmission of the disease. If they had done so it would have been a pleasure to have received their letters.

The statistics that are drawn from this correspondence are founded entirely upon what is given here, and are open to any criticism that may be directed against them by reason of personal opinion, or facts that can be brought against them by other observers. It is acknowledged that they are not absolute, and that, being limited to the one point spoken of, they do not show many others that would be of interest. It may also be that many of the gentlemen who replied in the negative might have given a positive reply if they supposed that the inquiry extended beyond cow's milk, for many of them specified this form of milk in their replies. It is, however, believed that, taken as a whole, there is much of value that may be drawn from a careful perusal and collation of the letters.

The correspondence follows; the letters from medical men are included between number 1 and number 168, whilst the replies of the veterinarians run from 169 to 180.

1.

No, neither to milk, nor other animal food.

B. F. D. Adams, Colorado Springs.

2.

Though I have made diligent inquiry for the past six years, in many cases, I have never once been able to trace a case of tuberculosis to a milk supply as a possible source.

Yours very truly,   John P. Bryson.

St. Louis.

3.

30th & Olive Sts., St. Louis, February 10, 1890.

Dear Doctor, — I have thought it worth while to do a little more than merely answer your inquiry in regard to the origin of tuberculosis in a milk supply. I ought to say that my experience of the disease is almost wholly confined to "Tuberculosis Uro-Genitalis." In the past six years I have studied a great many of these cases, and in the great majority the respiratory organs were found by me, and others who aided me, to be free of the disease, — in the beginning at any rate, — not being affected except in the latest stages, and sometimes, rarely, not at all. In all my carefully studied cases, the disease seemed to have reached the organs through the hæmatic channels. In my case-book, there is recorded one case where the disease began, apparently, in the left testis as three tubercular nodules. The patient (married and aged 36) declared to me that he had never taken any milk into his stomach since he was a child, — the thought of it making him sick. My very great interest in the subject has prompted this note, and I am pleased to see that the source of infection is being studied.

Very truly yours,
JOHN P. BRYSON.

To Dr. HAROLD C. ERNST, Boston.

4. (*Reply to above.*)

BOSTON, February 14, 1890.

MY DEAR DOCTOR, — Accept my thanks for your very kind reply to my circular letter in regard to tuberculosis and milk. I hope that I shall be able to get together some sort of basis for action in regard to controlling the use of milk from tuberculous cows.   Very truly yours,

HAROLD C. ERNST.

To JOHN P. BRYSON, Esq., M. D., St. Louis.

5.

I have no positive knowledge of such a case as you refer to in your circular.

J. BYRNE, Brooklyn, N. Y.

March 17, 1890.

6.

Dr. H. C. ERNST.

DEAR SIR, — In reply to your inquiry as to whether I have ever seen a case of tuberculosis which I have been able to trace to a milk supply as a cause, I can only say that I am unable to give you any information, having never made such an inquiry. I am personally much interested in the subject, as my breakfast for the past twenty years has been a bowl of bread and milk with a cup of coffee. So far, I have escaped contagion. Your inquiry is a very important one, and I shall keep the subject in mind carefully in the future. In the meanwhile I shall continue my usual breakfast.

Yours,   A. B. BALL.

7. (*Reply to above.*)

BOSTON, February 8, 1890.

MY DEAR DOCTOR, — I am obliged to you for your personal note in reply to my letter asking for information in regard to tuberculosis and milk. The matter seems to me to be one of extreme importance, but the last thing I desire to be considered is an "alarmist." The evidence to be derived from this letter is to be used in connection with certain experimental evidence . . . in order to an attempt to obtain a restriction of the sale of milk from tuberculous cows. . . .Your expression of interest is my apology for intruding upon you again.   Very truly yours,   HAROLD C. ERNST.

A. B. BALL, Esq., M. D.

8.

February 10, 1890.

Never saw such a case of tuberculosis that I can feel certain of, and, since we are acquiring the bacilli through the air so abundantly and constantly, don't think we need fear the milk source of infection greatly. The fact is, those who can, kill the bacilli, however acquired, — those who cannot are killed by them.   N. BRIDGE, Chicago.

9.

WOBURN, February 20, 1890.

Dr. H. C. ERNST.

DEAR DOCTOR, — In answer to your circular letter, I would say that in two instances I thought that tuberculosis was due to the milk supply, but I was unable to make such a connection as was satisfactory even to myself.

Yours truly, GEO. P. BARTLETT.

10. (*Reply to above.*)

BOSTON, February 23, 1890.

MY DEAR DOCTOR, — Will you not be good enough to give me the details in regard to the two cases where you suspected the origin of tuberculosis from milk? I want suspicious cases as well as those where the facts are perfectly plain, because the sentiment of the medical profession is as important as reports of cases. I hope that you will feel like doing what I ask. Very truly yours,

HAROLD C. ERNST, Harvard Medical School.

Dr. GEO. P. BARTLETT.

(*To this no answer was received.*)

11.

DEAR DOCTOR, — I cannot answer to your note as fully as I would like to, for the reason that I do not know whether it relates to a too great or too scant supply. But I will say that in all my cases, which have come under my care in midwifery, which have been quite numerous, — 3400, — I have, during fifty years' practice, had only two which were so distinctly marked as to give me the utmost assurance that first, they were well-developed cases of tuberculosis, and secondly they could be distinctly traced to almost a total lack of milk secretion. I have intended to arrange my cases by classification, but I am so feeble that I could not undertake the task. I may be able to give you a fuller synopsis.

Fraternally,

D. HOWE BATCHELDER, Danversport, Mass.

12.

DEAR DOCTOR, — In reply to your circular, I cannot say that I have ever seen a case of tuberculosis which it seemed possible to trace to a milk supply as a cause; but I do not doubt that the milk supply may be an important factor in the production of the above disease in some cases.

ANDREW BAYLIES, M. D.

13.

TURNER'S FALLS, MASS., February 19, 1890.

DEAR DOCTOR, — In answer to yours of recent date, will say have never had a case when it was possible to prove it due to the milk supply, yet I firmly believe it was.

Hastily yours,   E. G. BEST.

14.

BARRE, MASS., February 10, 1890.

H. C. ERNST, M. D.

SIR, — In answer to your inquiry, I would say, that I have never had a case of tuberculosis that I could trace to a milk supply. Several years since there was a cow in this town that evidently had tuberculosis, but the milk or beef was not used. No post-mortem.   Respectfully,

L. F. BILLINGS, M. D.

15.

SHERBORN, MASS., February 8, 1890.

DEAR DOCTOR, — I do not think I ever had a case of tuberculosis which I could trace to a milk supply as a cause. It should be said, however, that this is an agricultural town, and a large part of the people make the milk which they use on their own premises. I would also add that the percentage of cases of that disease is much smaller now in this town than it was twenty-five or thirty years ago.

Yours truly,   A. H. BLANCHARD, M. D.

16. (*Answer to above.*)

BOSTON, February 10, 1890.

MY DEAR DOCTOR, — Thank you for your reply to my letter in regard to tuberculosis and milk. May I ask you whether the diminution in your town of the disease (tuberculosis) is a matter of personal observation or of record? It is an interesting fact to know. Very truly yours,

HAROLD C. ERNST.

A. H. BLANCHARD, M. D., Sherborn, Mass.

17. (*Reply.*)

SHERBORN, MASS., February 15, 1890.

DEAR DOCTOR, — Referring to your letter of February 10th, in which you ask whether the decline in the number of cases of tuberculosis is a matter of personal observation or of record, I reply that it is both. It is from personal observation since 1851, and from record since 1841, when the cause of death was first recorded in our town register. For about fifteen years, from 1841, the cases of death from that disease were fully 25 per cent. of the whole number of deaths. Since that time there has been a gradual diminution, until in the ten years, 1880–89, the rate has been but 8.5 per cent. of the total number. Those figures are obtained from the town record. I think during the latter years there has been less of overwork among farmers and their wives, and that there has been generally a more careful observance of the laws for the preservation of health ; and this may have had something to do with the decrease in that disease.

Very truly yours, A. H. BLANCHARD.

H. C. ERNST, M. D.

18.

MY DEAR DOCTOR, — Your circular duly received, and evidently became mislaid. Am not aware of any case ever coming under my observation which could be consistently ascribed to milk supply. It would seem to be a question that

ought to be settled, — the possibility of infection from milk
supply, — and make our precautions conform thereto.   Success to your labors.   Yours cordially,

A. G. BLODGETT.

WEST BROOKFIELD, MASS., July 9, 1890.

19.

February 6, 1890.

DEAR DOCTOR ERNST, — I have often thought of the milk
supply as a source of tuberculous infection, and have sought
to connect it with the disease, but have *not* succeeded in so
doing.   I am glad that you are doing this work, which is, in
my opinion, of great importance.

Yours truly,

ALBERT N. BLODGETT.

P. S. — I suppose you mean cow's milk.

20. (*Of inquiry to preceding.*)

BOSTON, February 8, 1890.

MY DEAR DOCTOR, — Thank you for your reply to my
letter.   Do you mean to imply that you have ever seen cases
that seemed to you to be due to nursing a tuberculous woman?
If so, I want all the details that you are inclined to give me,
if you will be good enough to send them to me.   The matter
seems to me to be one that requires immediate and thorough
investigation, and I suppose that it is needless to refer you to
the experimental evidence that I have offered in a recent
number of the "American Journal of Medical Sciences."

Sincerely yours,   HAROLD C. ERNST.

21. (*Answer to* 20.)

BOSTON, February 10, 1890.

MY DEAR DOCTOR ERNST, — Yours of the 8th at hand,
and I am sure you will pardon me for the allusion to mother's
milk, which I made only because I know of your good work
in this important direction.   I have at present no definite

results to communicate, but I have taken some observations which are not at present in a form to communicate, but which I will take the liberty to present if they reach any practical form. I have followed your communications in the " American Journal " with much interest, and think we all owe you a debt of gratitude for your painstaking labors in a direction which presents peculiar and almost insurmountable obstacles to the investigator.   Yours sincerely,

ALBERT N. BLODGETT.

22.

I think not, as we have an abundant supply of pure milk.   J. M. BLOOD.

ASHBY, MASS.

23.

NEWTON CENTRE, February 11, 1890.

DEAR DOCTOR, — I have never met with a case of tuberculosis that I could directly trace to a milk supply.

Yours,   J. H. BODGE.

24. (*Of inquiry to preceding.*)

BOSTON, February 14, 1890.

MY DEAR DOCTOR, — In your reply to my circular in regard to tuberculosis and milk, you seemed to imply that you had heard of such a case as was there inquired about. If that is so, will you not be good enough to send me the account of it, or put me in the way of getting such an account? The importance of the subject is my excuse for intruding upon you again.

Very truly yours,

HAROLD C. ERNST.

Dr. J. H. BODGE.

No reply was ever received to the above.

25.

113 BOYLSTON STREET, February 7, 1890.

DEAR DOCTOR, — I have never seen a case which I thought attributable to milk. You are engaged in a most important work, and I wish I could help you more than by a simple negative.   Yours truly,

HENRY I. BOWDITCH.

26.

DEAR DOCTOR ERNST, — I have no positive data bearing upon the question.   Yours,   W. P. BOWERS.

27. (*Of inquiry to preceding.*)

BOSTON, February 14, 1890.

DEAR DOCTOR, — Even if you have no positive evidence in regard to the communication of tuberculosis by the milk supply, will you not send me any suspicious cases that have come under your observation?

Sincerely yours,

HAROLD C. ERNST.

Dr. W. F. BOWERS, CLINTON, MASS.

No reply received to the above.

28.

LOWELL, MASS., February 10, 1890.

Dr. H. C. ERNST.

DEAR SIR, — I certainly never did see a case as designated. I lived and practiced over eight years in and near Montreal, D. C., and here over eight years too, and I have not yet seen a case that I could surely and beyond peradventure trace to the milk supply. And yet I have been a man of observation in that direction for a purpose.

Yours very truly,

H. R. BRISSETT, M. D.

29. (Of inquiry to preceding.)

BOSTON, February 12, 1890.

MY DEAR DOCTOR, — Would you be willing to give me any details in regard to cases that have come under your observation where you even suspected the milk as a cause of the transmission of tuberculosis? Any evidence is of value in such a matter as this.   Very truly yours,

HAROLD C. ERNST.

H. R. BRISSETT, M. D., Lowell.

30. (Reply to preceding.)

LOWELL, February 25, 1890.

Dr. ERNST.

DEAR SIR, — I have carefully looked through all notes that I possess dating twenty years back, and cannot find nor recall a single case that I could trace to tuberculous infection from the cow, nor can I recall one case of tuberculosis that I even remotely suspected was of that origin, and so must dismiss the question with some sorrow at not being able to shed some light (faint even) on the subject, — a most important one.

Yours very truly,   H. R. BRISSETT, M. D.

A note of thanks was returned for the above.

31.

MY DEAR DOCTOR ERNST, — I have never encountered a tuberculous case which seemed traceable to the lower animals, although I deem such contagion quite possible.

Very truly,   W. E. BROWN, Gilbertville.

February 13, 1890.

32.

STONEHAM, MASS., February 10, 1890.

DEAR DOCTOR, — I have not met with any case of tuberculosis which could be traced to a milk supply.   Dr. Clarke, of Melrose, has met with several cases, and could furnish you with particulars.   Respectfully,

W. S. BROWN, M. D.

A letter of inquiry to Dr. Clarke, who had already sent a negative to the circular, called out the following : —

33.

MELROSE, February 15, 1890.

MY DEAR DOCTOR, — Doctor Brown is in error when he spoke of my paper, read before the Society. It was on *Diphtheria* and milk supply. I would be pleased to be of service to you.   Yours very truly,   J. S. CLARKE.

34.

I am unable to give you any information that would be of service to you. The milk supply is not by any means what we could wish, yet at the same time, whatever harm it may do is hard to state. In all cases of tuberculosis which I have observed, all conditions were such that the source could not be told, and I could not attribute to milk more than other foods or influences from external conditions.

Yours,   F. L. BURT.

751 TREMONT STREET, BOSTON.

35.

"No, from animal to man ; — yes, from animal to animal."   W. J. COATES, M. D.

141 W. 54TH STREET, N. Y.

36.

A letter of inquiry brought back the following : —

NEW YORK, February 21, 1890.

DEAR DOCTOR, — I deferred answering your note on account of sickness in my family. The cases I refer to were two calves, from different mothers, which were healthy and fed on milk from a tuberculous cow (Jersey). I had both cow and calves destroyed, and on post-mortem revealed tubercular deposits in lungs and other portions of the body. One of our cats was fed with milk from a cow suffering from tuberculosis, and developed the symptoms of phthisis, she becoming so emaciated and weak that she was destroyed after a period of eight or nine months, and showed tubercular deposits over the entire body. Other cats were fed on tubercular milk, and some developed the same. In regard to the human subject, it is difficult to trace, as there is too great a difference between cause and effect; by the time the physician could recognize the disease, the milk source would be lost sight of. A man might be ailing for many years, and his disease not appreciated by his physician, until some day he takes what is commonly termed a cold and develops acute symptoms of phthisis which will be given credit to atmospheric influences and not to a source of meat or milk supply which may have been years before. The milk question will probably not be settled.   Yours, etc.,

W. J. Coates, M. D., V. S.

A letter of thanks was returned for the above.

37.

Dear Sir, — I have never seen a case of tuberculosis which it seemed possible to me to trace to the milk supply as a cause. You are not to take this, however, as an expression of disbelief. At present it seems probable to me that milk may be a cause of tuberculosis.

Yours truly,

D. M. Cammann, M. D.

19 E. 33d Street, N. Y., February 8, 1890.

38.

920 Market Street, San Francisco,
February 13, 1890.

Dear Doctor, — My answer to your question is no. As bearing on the subject I may mention that I have never seen elsewhere tuberculosis so prevalent and deadly as it was in Alaska during my stay there — 1865–1872 — among the Indians, who had no milk except the human variety.

Geo. Chismire.

39.

A letter of inquiry to Dr. Chismire brought the following:

920 MARKET STREET, SAN FRANCISCO,
March 1, 1890.

MY DEAR DOCTOR, — I am sure I have seen statements of the prevalence of tuberculosis among the northern Indians in print, but for my life I cannot tell where. I would suggest your writing to the elder Dr. Helmican, of Victoria, British Columbia; he is a most competent man, and has had more than fifty years' experience while in the service of the Hon. Hudson's Bay Co.   Very truly yours,

GEO. CHISMIRE.

A letter of thanks was sent, but a note sent to Dr. Helmican met with no response.

40.

TOWNSEND, MASS., March 6, 1890.

DEAR DOCTOR, — I had occasion as an official of the board of health to condemn last summer a cow with consumption, and ordered killed; have seen no ill results from the use of the milk.   Yours,   L. G. CHANDLER.

A request for information of any facts turning up in connection with this case has as yet met with no response.

41.

Very few cases of tuberculosis in this neighborhood during the time I have known it.   C. A. CHEEVER,

Mattapan, Mass.

42.

65 CHANDLER STREET, BOSTON, February 10, 1890.

DEAR DOCTOR, — In answer to your circular, I can say that I do not think I have. But I fully believe that such transmission is possible, and think that too much care cannot be taken to guard against such impure milk.

Very truly,   E. CHENERY.

43.

BOSTON, February 10, 1890.

DEAR DOCTOR, — Your inquiry received. I think the query a very important one. My cases, however, have not led me to suspect milk as a probable cause, therefore to your question I must answer no.   Very truly,

C. H. COBB.

44.

ROXBURY, February 25, 1890.

Dr. ERNST.

DEAR SIR, — In reply to your circular letter will say that I saw a case that seemed to me might possibly be tuberculosis in the baby — no family history of tuberculosis — from the cow's milk. The baby was fed on a Jersey cow's milk (uncooked). It never prospered, lost flesh, developed a bronchitis and large belly, much swollen. I could not find enlarged glands, but otherwise it seemed like Tabes Mesenterica. In the mean while the cow was taken sick and died, and the baby died soon after, but another M. D. was in at the finish, and I did not know about it, to get a post-mortem.

A. B. COFFIN.

A letter of thanks was sent for the above.

45.

BOSTON, February 17, 1890.

I have seen two cases, both children, when it seemed possible to believe the milk was the primary cause. Babcock examined the milk, and as it was much below standard, the man was arrested.   W. M. CONANT.

A request for further information in regard to these cases was not responded to.

46.

BOSTON, February 16, 1890.

DEAR DOCTOR, — It is perhaps a fact worth mentioning that very few of the children under the care of the Department Outdoor Poor succumb to diarrhœal or digestive diseases. We often receive marasmic children.

Yours,   S. M. CRAWFORD.

47.

DEAR DOCTOR, — I do not think that I have met with a well-substantiated case of tuberculosis traceable to milk.

Very truly yours,

EDW. L. DUER.

PHILADELPHIA, 1606 LOCUST STREET,
February 19, 1890.

48.

A letter of inquiry as to whether Dr. Duer had seen any cases in which he had had reason to *suspect* such an origin of tuberculosis met with no response.

49.

PORTLAND, ME., February 20, 1890.

MY DEAR DOCTOR, — I have never been able to trace a case of tuberculosis to a milk supply, though I have repeatedly suspected the milk.   Yours very truly,

ISRAEL T. DANA.

50.

Having written to Dr. Dana for any suspicious cases that he might have, he replies as follows : —

PORTLAND, ME., February 20, 1890.

DEAR DR. ERNST, — I am afraid in my hasty letter in response to your circular I gave rather a wrong impression. It was rather a general impression, taken from observation, than any record of individual cases to which I referred. I have had cases of infants brought up on cow's milk, where neither heredity nor environment would lead to the expectation of tuberculosis, in which tuberculous symptoms have rapidly developed, with fatal terminations. The symptoms have oftener been abdominal than pulmonary. There have been frequent loose, ill-smelling dejections and general marasmus. The abdomen has been tumid and tender, sometimes giving to the touch the sensation, through the attenuated abdominal walls, of swollen mesenteric glands. In some of the cases the most natural explanation of the phenomena present has seemed to me to be in the line of infectious tubercle-producing cow's milk.   Yours very truly,

ISRAEL T. DANA.

51.

In reply to a query to that effect, Dr. Dana says that he was not able in any case to " push investigations so far as to ascertain that the milk supply came from a tuberculous cow.

51 a.

Not from cow's milk. Have seen an apparently non-tuberculous baby (waif) die, after nursing a few months from a tuberculous foster-mother, from tuberculosis.

Yours very truly,   F. F. DOGGETT.

A note was sent to Dr. Doggett, asking for details of this case with the following result.

51 b.

805 BROADWAY, BOSTON, February 11, 1890.

DEAR DOCTOR, — Yours of the 8th inst. received in regard to case. The case occurred on Athens street in District 9 of

Boston Dispensary while I was physician to that district.
The case so impressed me at the time that I made notes of it,
but am sorry to say that I have been unable to find them
after careful search.   However, if the few details of the case
which remain in my memory are of any service to you, you
are welcome to them.   The child was illegitimate, — was
plump and healthy at birth, — was deserted by its mother at
about six weeks old, and died of acute miliary tuberculosis
at about three and a half months.   When the child was
deserted, the foster-mother, who had just lost her own baby,
having milk in her breasts and pity in her heart, adopted the
waif.   I saw the child about six weeks after, when it was
three months old.   It was then greatly emaciated, with ascites
prominent, much diarrhœa, and signs of consolidation at both
apices.   The foster-mother's milk was thin and poor, but
quite abundant.   She had well marked phthisis, as I noted
on my dispensary book, — the details I know nothing about.
What her own baby died of I was unable to learn.   The
reputed father of the child was said to be in good health;
also the mother, who was a servant girl, and was working
when I first saw the child.   There was no autopsy.   There
were convulsions toward the last.

The hygienic surroundings were about as bad as they could
be, and it impressed me at the time that, excluding the milk
as the medium of infection, the sputa, which was abundant
from the woman and very carelessly disposed of, might well
be blamed.   Without autopsy it would perhaps have been
more exact to say pulmonary and abdominal tuberculosis
rather than general tuberculosis.

Yours very truly,   F. F. DOGGETT.

51 c.

Dr. J. R. Deane of Newton Highlands, Mass., returned
the circular letter, endorsed "Yes," but an appeal to him
(letter-book p. 281) for details of the case, or cases, that he
had seen received no reply.

51 *d*.

E. S. Dodge (Natick, Mass.) replies, " Emphatically, No."

52.

DEAR DOCTOR, — I must answer " No " to your query, but as I, like most physicians, can know so little of the milk supply of our patients it seems to me that negative testimony can have very little value. In fact, I do not see how this inquiry can lead to definite results in cities. In the country, where the doctor knows not only the families but often their beasts, one might come at positive results.

Yours very truly, R. T. EDES.

53.

HYDE PARK.

DEAR DOCTOR, — No, I have not, except from a tubercular mother. Truly yours, C. L. EDWARDS, M. D.

A letter of inquiry in regard to the above received no answer. It afterwards appeared that Dr. Edwards was ill.

54.

DEAR Dr. ERNST, — Have never seen a case which could be traced to milk with any degree of probability.

Very truly yours, E. P. ELLIOT.

DANVERS, February 11, 1890.

55.

A note asking for any cases where suspicion had rested upon the milk received the following reply.

DANVERS LUNATIC HOSPITAL, DANVERS, MASS.,
February 22, 1890.

H. C. ERNST, M. D., BOSTON, MASS.

DEAR SIR, — Your letter of the 14th, addressed to Dr.

Elliot, is received at this hospital. As Dr. Elliot has gone to Europe and cannot answer you at once, I volunteer to make the statement which your letter seems to call for. For a period covering the past five years the ratio of deaths from phthisis to the whole number of deaths in this hospital is about ten per cent. This is, I suppose, considerably lower than the average ratio for the whole State. This ratio varies but slightly, however, in the several state hospitals for the insane, viz. : Worcester, Westboro, Taunton, and South Boston. While these hospitals maintain a large population of chronic patients they are constantly receiving new cases. At the Northampton Hospital, where but a comparatively small number of men patients enter, the ratio for the same time, the last five years, is above twenty per cent.

We had eleven deaths from phthisis and one from Bright's disease and phthisis during our last hospital year. Of these twelve patients, eight were cases of chronic insanity, three were cases of acute melancholia, and one had general paralysis. I believe you are engaged in a most important study and should be glad to assist you in establishing the facts, but I cannot discover a " scrap of evidence " at Danvers that milk causes tuberculosis.   Very respectfully,

CHAS. W. PAGE,
Physician and Superintendent.

A note of thanks was returned for the above.

56.

I think that I have, but as it is matter of opinion and not of demonstration, I am unable to make any observations upon it that are of any value. I am very glad that you have started in this work. I wish there were some method by which I could aid in it, for it seems to me to be very closely connected with the public health. It has been uppermost in my mind for years, but I did not feel myself equal to taking hold of it, and am only too happy that its importance is in a fair way of being demonstrated.

Very truly yours,   W. S. EVERETT.

HYDE PARK, February 15, 1890.

An appeal for details of cases where the writer suspected the milk supply as a cause of tuberculosis failed to meet with a response.

57.

CINCINNATI, February 17, 1890.

DEAR SIR, — I have never seen a case of tuberculosis in which a positive connection could be established between it and tuberculosis in the cow (milk supply). Although not within scope of your question, I am convinced of the fact that such connection does exist.   Yours truly,

F. FORCHHEIMER.

Dr. HAROLD C. ERNST.

58.

A letter was sent to Dr. Forchheimer, asking if he had seen cases where the suspicion of the origin of tuberculosis from milk had been aroused in his mind. This was replied to as follows : —

CINCINNATI, February 19, 1890.

DEAR DOCTOR, — In reply to yours of the 12th, I would state as follows : according to my notion, tuberculosis is *by far* the most common of children's affections, — again, most common in a localized form. The place where it is most frequently found in them is somewhere in the alimentary tract or organs connected with it. Milk is the most common article of diet in children ; milk contains tuberculous material to an extent which, according to my idea, is not properly estimated, so that I have the conviction that tuberculosis is frequently caused by milk. As to a record of cases of this connection, or scientific proof of the same, I should hesitate a very long time before I would put down any individual case as in evidence. Cases are not uncommon, in practice, in which a tuberculous mother nurses an infant which dies, let us say, of a meningitis tuberculosa. Yet, in such a case, in which I am convinced that the mother has transmitted tuberculosis to her child, how can I present evidence sufficiently conclusive to prove that the infection has not come from another, extraneous source? I have seen children who, according to the statement made to me, have had no other food but milk, with the following set of lesions : tuberculosis of the glands about the neck, of intestine, mesenteric glands, lungs, and meninges. I am justified, I think, in the conclusion that the tuberculosis was produced by a something introduced into the alimentary canal. I am convinced that it was by means of milk, yet I am not justified in this individual case in stating that this was the cause to my knowledge. In other words, I cannot put down such a case as one capable of exact demonstration.

I hope I make my meaning clear. If such cases as I have referred to will be of any service to you, I will be very glad to hunt through my records for you.

Very truly yours,   F. FORCHHEIMER.
Dr. HAROLD C. ERNST.

A note of thanks was returned for the above.

59.

I have never yet seen a case of tuberculosis that I felt could be laid to milk supply, unless it was a mother's milk. I think that our milk supply is good, and from well-managed farms and good healthy cows.   Respectfully yours,

U. H. FLAGG, M. D., Mittineague, Mass.

A letter of inquiry for definite information in regard to any cases of transmission by means of mother's milk failed to call out a response.

60.

283 ESSEX ST., LAWRENCE, MASS., February 8, 1890.

DEAR DOCTOR, — No, I never did ; but the possibility of tuberculosis in the udder of a cow being propagated to the

human race has for the last five years been a source of un-
easiness to me.   Yours sincerely,

F. B. FLANDERS.

61.

ANN ARBOR, MICH., February 9, 1890.

Not of tuberculosis, but several of Tabes Mesenterica.

HENEAGE GIBBES.

A request was sent for information in regard to the cases
of Tabes Mesenterica spoken of above, and called forth the
following : —

62.

UNIVERSITY OF MICHIGAN, PATHOLOGICAL LABORATORY,
ANN ARBOR, February 17, 1890.

MY DEAR SIR, — I am under the impression that your
views on tuberculosis and mine are opposed. I am now writing a paper on this subject, in which I shall utilize the cases
I mentioned. I think it would not do for the same cases to
appear on opposite sides of the same subject, otherwise I
should have gladly sent you an account of them. I am

Yours very truly,
HENEAGE GIBBES.

Dr. HAROLD C. ERNST, Boston.

63.

The following reply was sent to the above.

BOSTON, February 23, 1890.

MY DEAR SIR, — I regret that you should feel that you
cannot send me an account of the cases that you spoke of. I
hope that I am not more stubborn of conviction than most
men ; and if I am not easily moved by striking evidence I
am unconscious of the fact. I hope that I shall see the paper
upon tuberculosis when it is published, and beg to apologize
for my indiscreet intrusion upon you.

Very truly yours,   HAROLD C. ERNST.

Dr. HENEAGE GIBBES, Ann Arbor, Michigan.

64.

In reply to the above note, Dr. Gibbes writes as follows : —

ANN ARBOR, February 25, 1890.

MY DEAR SIR, — I am afraid I expressed myself badly in my last letter to you. I have read several of your papers, and gather from them that you are convinced the tubercle bacillus is the cause of tuberculosis ; from this I conclude you consider human and bovine tuberculosis identical. Now I am not satisfied on these points, and intend to use the cases I have in support of my argument. Your circular and letter gave me the idea you were collecting evidence in support of your views, and I could not give you my facts for your side of the question.   Yours very truly,

HENEAGE GIBBES.

To the above no reply was sent, although it might easily have been said that *facts* are the same whichever side they are used upon.

65.

LOWELL, March 3, 1890.

DEAR DOCTOR ERNST, — The accompanying sample of milk is from a cow that has furnished milk to a child now suffering from meningitis (whether tuberculous or not I am not yet sure). I have had the cow examined by a veterinary surgeon. He says that the lungs sound rather suspicious, but the symptoms are not yet characteristic. I thought you would be interested to look over a sample of "strippings" for bacilli. I will let you know the outcome of case.

Very sincerely yours,

J. ARTHUR GAGE, 48 Central St.

66.

Shortly afterwards a second letter came from Dr. Gage, as follows : —

DEAR DOCTOR, — I sent you recently a specimen of milk, and I write now to tell you that the child died yesterday. Although no autopsy was obtained, the symptoms and course were sufficiently distinctive to warrant a diagnosis of tubercular meningitis. The family history is good, and the food consisted (aside from breast milk) only of milk from one cow. I shall keep the cow under surveillance, and shall be glad to hear whether you found any bacilli in the milk. Provided you would like another specimen, I will procure and send you one. Very sincerely yours,

J. ARTHUR GAGE.

Dr. HAROLD C. ERNST.

67. (*Reply to the above letters.*)

BOSTON, March 15, 1890.

MY DEAR DOCTOR, — I am in receipt of your letters, and should have answered them before this, but that I have been overwhelmed with work. I got the specimen of milk all right, and used it for inoculation and cover-glass experiments. If there is any result I shall be glad to let you know; in the mean time please accept my thanks for your kindness and the trouble that you have taken. Very sincerely yours,

HAROLD C. ERNST.

J. ARTHUR GAGE, M. D., Lowell.

The result of this inoculation is given in its proper place, and was the death of three out of four of the rabbits inoculated, as shown in the record of experimental work. Early in May a note was sent to Dr. Gage telling him of the result of the inoculations, and asking if one of my assistants could see the cow if he came to Lowell. The reply is below.

68.

LOWELL, May 8, 1890.

DEAR DOCTOR ERNST, — Your letter just at hand. I reply at once to say that I will fill out records for you, and would

like to have some one come up to look over the cow with me.
I understand that another child has been fed on the same
cow's milk, and I will look up the matter.

Yours very truly,   J. Arthur Gage.

The records, as promised above, have never come, and one
of my assistants (Dr. Frothingham) went to Lowell, and was
unable to find any definite signs of tuberculosis in the suspected animal.

69.

Boston, February 7, 1890.

To mother's milk, yes; to cow's or other domestic animals,
no.

George W. Galvin.

United States Hotel.

A letter of inquiry was sent to Dr. Galvin, with the following result : —

70.

Boston, February 11, 1890.

H. C. Ernst, M. D.

Dear Sir, — The only case to which I can refer you is at
13 Edinboro Street. Ask for Mr. Clark. I have had several cases which, to my mind, were tuberculous, through the
mother's milk. I may be able to furnish one more as soon as
I ascertain the condition of the child. I told Mr. Clark to
expect you or your assistant. Very truly yours,

George W. Galvin.

A letter of thanks for this note was returned, and Dr.
Jackson visited the family, sending in the following note of
the case : —

"A boy six years old, — tuberculous. Nursed by his
mother, who, while nursing him, developed a cough, and died
three years later of pulmonary tuberculosis."

71.

Never, never, never. It may be possible, but not probable, unless the cow has been dissipated, and a free user of alcoholic drinks!

T. GARCEAU.

ROXBURY.

72.

SPRINGFIELD, February 8, 1890.

Dr. H. C. ERNST.

DEAR SIR, — In answer to your printed query, *No!* (I understand your germ theory, with milk for a vehicle.) But I have a case of a woman in my own family who has chronic tuberculosis, and who had "la grippe," followed by pneumonia, or second stage, and *who took no medicine* but milk, constantly sipping it night and day, according to her whim. She is able to be about the house, and is better than before the "grippe." She is seventy-three years old.

Yours,   W. W. GARDNER.

73.

GLOUCESTER, February 15, 1890.

DEAR DOCTOR, — In answer to your circular I would say that I have had no case which could be satisfactorily traced to a milk supply as a cause. Yours truly,

A. S. GARLAND.

A note of inquiry in regard to any cases where suspicion was aroused received no reply.

74.

QUINCY, MASS., February 7, 1890.

DEAR DOCTOR, — A child about ten months, bottle-fed, developed tuberculosis and died. The cow from which the milk was obtained died of tuberculosis a few weeks afterwards. Yours very truly,

J. A. GORDON, M. D.

A request to Dr. Gordon for any further details in regard to the case mentioned above, brought this reply : —

75.

March 21, 1890.

MY DEAR DOCTOR ERNST, — In reply to your note of February 10, relating to the question of tuberculosis and the milk supply, I am very sorry to say that I have no notes of the case I mentioned, although my memory serves me fairly well as to the main facts, which are as follows : A child of perfectly healthy parents, with no hereditary or present history of consumption, wasted and died with symptoms pointing unmistakably to tuberculous disease. After the death of the child I ascertained that the cow which had supplied the milk, which had been the exclusive diet of the child for several months, had had a cough for some time, and died with signs and symptoms of pulmonary tuberculosis a few weeks afterwards.   Yours very truly,

J. A. GORDON.

A note of thanks was returned to Dr. Gordon for his letter.

75 a.

Since the relation between tuberculosis and milk has been under discussion, I have had little or no experience with tubercular disease, but my answer to the question proposed is No.

R. M. HODGES.

408 BEACON STREET, BOSTON.

A letter of thanks was sent to Dr. Hodges, with a request for any suspicious cases of tuberculosis coming in infants from nursing tuberculous women, that he might have seen. He replied as follows : —

75 b.

408 BEACON STREET, February 10, 1890.

DEAR DOCTOR ERNST, — I have never had "reason to suspect the occurrence of tuberculosis in an infant after, and because of, nursing a tuberculous mother." My experience is small as to families in which there have been children of tuberculous mothers whom I have had for long periods under my care or observation. I have always forbidden nursing where I suspected tuberculosis in the mother. I have always scrupulously stopped the nursing of babies by wetnurses with a cough, and have done this on general principles, which I suppose must have led other physicians to do the same thing. Did you get Dr. Morrill Wyman's opinion on the question ? Yours sincerely,

R. M. HODGES.

A note of thanks was returned to Dr. Hodges for his letter.

76.

February 11, 1890.

DEAR DOCTOR, — I have not seen personally a case where I thought tuberculosis was traceable to milk supply as a cause. Regretting that I am unable to assist you, I am
Yours respectfully, E. E. HOLT.

PORTLAND.

A note was sent to Dr. Holt asking for reference to any one who had seen such a case, and he kindly sent the following : —

77.

Dr. Geo. H. Bailey will give you details of cases.

Upon writing to Dr. Bailey for any information he might be able and willing to give, he replied by the following letter : —

78.

CATTLE COMMISSIONER'S OFFICE,
STATE VETERINARY SURGEON,
PORTLAND, February 15, 1890.

MY DEAR DOCTOR, — Yours of the eighth was received during my absence from home. In answer to your inquiry, "If I have ever seen a case of tuberculosis which it seemed possible to me to trace to milk supply as a cause," I feel perfectly warranted in answering "yes." I have a case now under observation where, about a year ago, I condemned a tuberculous cow, that proved upon post-mortem to be an advanced case of pulmonary tuberculosis. The milk from this cow was the sole supply of the family (a man and his wife), and although there is no history in the family of the woman that can possibly be traced to phthisis, she is in an advanced stage of consumption, as I have every reason to believe from the direct use of the milk of the cow that I condemned. I have had another case that closely approximates to the above, but where the history involves the grandparents of the subject. I send you my report of 1888, although I suppose the subject treated on pages 10, 11, 12, 13 are perfectly familiar to you. I am very truly yours,

GEO. H. BAILEY, D. V. S.,
State Veterinary Surgeon.

A note of thanks was returned to Dr. Bailey for his letter and report.

79.

NEWARK, N. J., February 7, 1890.

DEAR SIR, — Have always believed in the infectious character of tuberculosis, and published a pamphlet and article in the "American Journal of the Medical Sciences" some years before Koch's discovery, and while I believe that milk from a tuberculous cow might infect, have never yet been able to demonstrate it.   Yours truly,

EDGAR HOLDEN, M. D.

A letter of thanks, with a request for a reference to his paper, was sent to Dr. Holden, but there was no reply.

80.

2. 7. '90.

No, I never have. I remember that Crookshank showed me a large tuberculous cow's udder at his laboratory at King's College, London, in June, '88, but cannot remember whether the milk had caused human tuberculosis or not; I think it had. Could you not write to him?

H. A. HARE, Philadelphia.

In accordance with the suggestion, a note was sent to Dr. Crookshank, and the following note was received from him.

81.

KING'S COLLEGE, LONDON, March 20, 1890.

DEAR SIR, — I have just returned from Egypt, and hasten to reply to your letter. I have *not* seen a case. I send you a copy of my report (Local Government Board), in which you will find information bearing upon this important subject.   Yours very truly,

EDGAR M. CROOKSHANK.

HAROLD C. ERNST, Esq.

Thanks were returned for this note, and for the report, which will be found mentioned in its proper place.

82.

PHILADELPHIA HOSPITAL.

DEAR DOCTOR, — I know of no case of tuberculosis traceable to infected milk. It seems to me that very exceptional opportunities for observation would be needed to enable one to detect such an origin of phthisis.   Yours truly,

F. P. HENRY.

PHILADELPHIA, February 7, 1890.
1635 Locust Street.

83.

DEAR SIR, — Only one case where it seemed possible to connect tuberculosis with a milk supply, — and this not conclusive.   Yours, etc.,   W. L. HALL.

A letter of inquiry for data in regard to this case failed to receive a reply from Dr. Hall.

84.

No, though we have tuberculous cows in many of our dairies. Have seen cases of tuberculosis which could only be caused by infection from some source. If you have not done so, would suggest that you write to Dr. A. H. Rose, U. S. Veterinary Surgeon at Littleton, Mass.

I am yours truly,

BENJ. H. HARTWELL.

AYER, MASS., February 7, 1890.

84 a.

PALMER, February 7, 1890.

DEAR DOCTOR, — No, I never have, and have never looked for anything of the kind, my especial attention never having been directed that way until my investigation of tuberculosis, etc., for material for paper read by me on Wednesday in Boston. I shall look out from this time.

Yours,   WM. HOLBROOK, M. D.

85.

HYANNIS, MASS., February 8, 1890.

HAROLD C. ERNST, M. D.

DEAR SIR, — Have not had a case traceable to the milk supply, though I have suspected it.

E. E. HAWES, M. D.

A letter of inquiry about cases where such an origin of the disease was suspected received no reply.

85 a.

Yes, one case.   Yours truly,   C. M. HULBERT.

SOUTH DENNIS, February 7, 1890.

Of course a letter was sent to Dr. Hulbert asking for details of this one case, but no reply was ever received.

85 b.

Dr. ERNST.

DEAR SIR, — Several suspicious cases have come under my observation, but opportunities for a full investigation were not afforded me ; therefore I am not sufficiently certain to be sure.   Yours truly,   W. H. HULL.

85 c.

NEWTONVILLE, MASS., February 8, 1890.

Dr. ERNST.

DEAR SIR, — Have never seen a case of tuberculosis that I could trace to a milk supply as a cause, and have never seen any evidence that tuberculosis could be communicated by contagion.   Very truly yours,

OTIS E. HUNT, M. D.

85 d.

Dr. E. D. Hutchinson, Westfield, Mass., writes, "After an active practice of fifteen years, most decidedly no."

86.

88 CHARLES STREET, February 7, 1890.

DEAR DOCTOR, — I regret I have no new statistics on the subject, yet from several cases I had become suspicious that tuberculosis originated in the child from nursing, and therefore have for a long time insisted that where the mother was suffering from tuberculous disease that the infant should be reared "by hand."   Very truly yours,

CHAS. E. INCHES, M. D.

A letter asking for more exact details from Dr. Inches called out the following reply: —

87.

88 CHARLES STREET, February 13, 1890.

DEAR DOCTOR, — I am sorry I can give you only my suspicions, and the consequent prohibition of nursing by tuberculous mothers. Of course it is probable that the tubercular disease in the infant may be hereditary, and not due to bacilli in the mother's milk. I have no records on the subject.   Very truly yours,

CHAS. E. INCHES, M. D.

Thanks were sent to Dr. Inches for the above.

88.

MY DEAR DOCTOR, — In reply to your circular letter I beg to say that while I do not know that the milk supply has been the cause of tuberculosis that has fallen under my care, neither do I know to the contrary. The bulk of our milk supply comes from a great distance and it would be difficult to trace a suspected milk to its source. Regretting that I cannot help you, believe me

Most truly yours,

CHAS. JENRETT, Brooklyn.

89.

DEAR SIR, — I have no facts that I could prove, but that one cannot expect from those who practice in large cities, where the source of the milk cannot be traced.

Very truly,   A. JACOBI.

A request for any cases where suspicion was aroused received no reply from Dr. Jacobi.

89 *a*.

MY DEAR DOCTOR, — In replying in the negative to your question, I desire to express my great interest in the subject which is engaging your attention. While firmly convinced that many cases of tuberculosis in children which I see, have their origin in infected milk, such a genesis is exceedingly difficult to demonstrate in a great city, with its milk supply drawn from so many sources. I have had some unpleasant experience with the prevalence of the disease in even the most carefully selected herds. Suspecting its presence in my brother-in-law's herd, the opinion was confirmed by killing the entire herd (11), and finding on autopsy, tuberculosis, in every degree of severity, in all its members. I am

Yours very truly, FRANCIS P. KINNICUTT.
42 W. 27TH ST., N. Y.

A note of thanks was sent to Dr. Kinnicutt for his letter.

90.

BELLEVUE, FLORIDA, February 10, 1890.
H. C. ERNST, M. D.

DEAR SIR, — Your inquiry at hand. I have never seen in my practice a case of tuberculosis that I thought I could trace to the milk supply, but I believe that such cases do occur.

Yours, C. H. KNIGHT.

90 *a*.

CHERAW, S. C., February 18, 1890.
HAROLD C. ERNST, M. D.

DEAR DOCTOR, — Your letter in reference to tuberculosis which it seemed possible to trace to a milk supply as a cause, has been received. In reply I beg to be allowed to report briefly the following case that came under my observation. On the tenth of April, 1869, Mrs. A. J. L., a strong healthy woman, in the higher walks of life, aged 22 years, gave birth to a strong, vigorous male child, weighing ten and one half pounds. This lady had always enjoyed excellent health, and there was never a single case of pulmonary disease known among her ancestors, on both sides for three generations. They were all long-lived people. She was very ill after her confinement, partly from some neglect or mismanagement during her lying-in, and never nursed her child a single time. I saw her in consultation on the eighteenth of April, eight days after her accouchement, and found her in a deplorable condition. . . . She died the next night, — nineteenth of April.

Mrs. R. C. W., aged nineteen years, gave birth to a dead infant on the eighth of April, two days prior to Mrs. A. J. L.'s confinement. Having an abundant supply of milk, Mrs. R. C. W. offered to nurse the child of her friend and near neighbor. I remonstrated against this, but with no effect. Mrs. R. C. W. was well advanced in pulmonary consumption. This I know positively, for I had examined her lungs, and prescribed for her from time to time. Both parents and two or three brothers of Mrs. R. C. W. had died of pulmonary disease. At the age of twenty months, this vigorous child of Mrs. A. J. L. began to pine and show signs of a want of thrift and vigor. At this age it had a troublesome cough, which continued with more or less severity till ten or eleven, when it had a hemorrhage. It had several hemorrhages and died in its fourteenth year. I made an autopsy of the child and found both lungs riddled with tuberculous deposits. Mrs. R. C. W. died when twenty-four years of age. I made an autopsy of her case, and found her lungs in a similar condition to those of her foster-child.

I pronounced this an undoubted case of tuberculosis being transmitted through the milk of the woman who nursed the child. . . . If these facts I have stated will be of any service to you, I shall be glad to know that I have aided you in your laudable work. If I can serve you further, please command me.   Very truly yours,
                              CORNELIUS KOLLOCK, M. D.

A letter of thanks was sent to Dr. Kollock for his interesting letter.

91.

402 Washington Ave., March 6, 1890.

. . .In reply to your question I would say that I have never seen myself a case of tuberculosis directly traceable to milk supply. Truly yours,

PAUL II. KETZSHMAR, Brooklyn.

91 b.

Yes, but not with scientific accuracy. There were three cases which came under my care from another. All died. No other cause tenable. J. A. KITE, Nantucket, Mass.

A letter to Dr. Kite asking for further details of these cases was not replied to in any way.

92.

Dr. G. King (Franklin, Mass.) writes: Never had any reason to think that milk was the cause, in the remotest degree, of tuberculosis.

93.

DEAR SIR, — In reply to your inquiry as to the causation of tuberculosis by milk, I would reply that I have never seen a case in which I have traced the connection. My practice, however, is a special one, and it is a rare thing for me to see these diseases at all. I see with great pleasure that you are interesting yourself in this important inquiry.

Yours very truly, BENJAMIN LEE.

A letter of thanks was returned to Dr. Lee for his expression of interest.

94.

To HAROLD C. ERNST, M. D.

I have never seen a case of tuberculosis with any proof that it was due to food conveyance, nor in which it seemed possible to trace to milk supply, — the best of all foods for those liable to phthisis.   Respectfully,

J. R. LEAMING, M. D.,

18 West 38th Street, N. Y.

95.

PHILADELPHIA, February 19, 1890.

I am not aware that I have ever seen a case of tuberculosis traceable to a milk supply as a cause. Of course this inquiry does not apply to a want of sufficient supply of milk to the infant, which is no doubt a frequent cause of disease.

JAMES J. LEVICK.

96.

CONWAY, February 7, 1890.

DEAR SIR, — I have not, but am on the watch, as two of our farmers have tuberculosis in their barns.

Yours truly,

DR. J. B. LAIDLEY.

A letter of thanks and request for any further information that might arise was sent to Dr. Laidley.

97.

DEAR DOCTOR, — I have never been able to trace a case of tuberculosis directly to the milk supply. It may be of some interest to you to know of a man that kept a cow within the city limits. She was tuberculous; gradually lost flesh until she was little else than skin and bones. Being fed on "brewers' grains," her milk was sufficient for two families. In one family there were two adult sons; both took this milk, both became sick, one or both are dead of consumption. Others saw the cow and would not take her milk; of these none were ill.   Yours truly,

HENRY F. LEONARD.

781 TREMONT STREET, BOSTON,
February 7, 1890.

A request for further information received the following reply : —

98.

781 TREMONT STREET, February 11, 1890.

DEAR DOCTOR, — Yours received. I regret to say I can write nothing more definite than you learned by the postal-card. My patient, the observer and informer, is now in California. The two consumptives are dead, and probably the cow also. I read with interest the articles you refer to. If possible to learn more of this case I will send you word later.

Yours sincerely,   HENRY F. LEONARD.

99.

HAVERHILL, February 17, 1890.

DEAR DOCTOR, — I don't think I quite understand what you mean by "a milk supply." If you mean a milk diet, I think I have ; if you mean something else, I don't know.

Respectfully,
OLIVER S. LOVEJOY, M. D.

A letter of inquiry and explanation was sent to Dr. Love-joy, but did not elicit a response.

100.

NEWTON HIGHLANDS, February 8, 1890.

In answer to your inquiry, I have seen no case of tuberculosis where it was evident that a milk supply was the cause.

J. D. LOVERING, M. D.

A letter asking for suspicious cases received no reply.

101.

DEAR DOCTOR, — A case of the kind has never come under my personal observation.

Truly,
· P. A. MORROW,
66 West 40th Street, New York.

A request for information concerning any cases occurring to others which Dr. Morrow might have heard of, received no reply.

102.

PHILADELPHIA, PA., 1417 WALNUT STREET,
February 10, 1890.

DEAR DOCTOR, — I know of no case of tuberculosis caused by the use of cow's milk in my individual experience.

Yours truly,   ALEX. W. MACCOY.

A letter asking for any information of cases *heard of* was not replied to.

103.

DEAR DOCTOR, — In answer to your circular note in regard to the transmission of tuberculosis from cow to man, I would say that I have never seen a case in which such transmission could be demonstrated beyond the possibility of doubt.   Yours very truly,

JOHN W. MACKENZIE.

February 17, 1890.

A request for information of any cases where suspicion was aroused of the causation of the disease by milk received this reply : —

104.

605 NORTH CHARLES STREET, March 3, 1890.

DEAR DOCTOR ERNST, — I would have answered your note sooner, but have been away from home and have not had the opportunity to do so. I am afraid that I misled you in my postal. The data, if such a term can be applied to them, in my possession are absolutely valueless in evidence, and I must therefore say that I have never seen a case in which there was reason to believe that the disease had been transmitted from one of the lower animals. Wishing you success in your researches,   Very sincerely yours,

JOHN W. MACKENZIE.

105.

DEAR DOCTOR, — I regret that I cannot give you any definite information from my experience. I have seen cases that might have come from milk supply, but under the present system of obtaining milk it would be impossible to prove anything.   Yours truly,

R. C. MACDONALD, M. D.

106.

TURNER'S FALLS, MASS., February 10, 1890.

Dr. H. C. ERNST.

DEAR SIR, — Would say in reply to your inquiry that I have never known personally a case of tuberculosis traceable to a milk supply.   Very truly yours,

C. C. MESSER.

A letter asking for any cases where the suspicion had been aroused in the minds of others, and which Dr. Messer had heard of, received the following reply: —

107.

TURNER'S FALLS, MASS., February 19, 1890.

HAROLD C. ERNST, M. D.

MY DEAR SIR, — Am sorry to say that my reply to your inquiry in regard to tuberculosis in milk evidently implied more than I intended. Have no definite knowledge upon the subject, only the general impression that it is so, obtained from reading and conversation. Wish that I could help you to some points upon the subject.

Very truly yours,   C. C. MESSER.

108.

HOLYOKE, February 13, 1890.

DEAR SIR, — In answer to your inquiry I must say that I never did see a case of tuberculosis which could be traced to milk, and I suppose I never will, for the question in the present status of our knowledge can no more be demonstrated than the presence of Connecticut River water in the middle of the Atlantic Ocean.

Respectfully yours,

W. W. MITIVIER.

To the above the following answer was returned : —

109.

BOSTON, February 16, 1890.

SIR, — I am obliged to you for your answer to my circular letter in regard to tuberculosis and milk, and regret that it does not meet with your approval. Permit me to remind you that no advance would ever be made in scientific subjects if one should hesitate to undertake the investigation of subjects as recondite even as the search for the waters of the Connecticut in the middle of the Atlantic Ocean.

Very truly yours, HAROLD C. ERNST.

110.

212 No. MAIN ST., PROVIDENCE, February 10, 1890.

Dr. H. C. ERNST.

DEAR SIR, — Have never seen a case of tuberculosis which could even approximately be traced to milk. It has fallen to my lot in the last three years to watch to a termination several cases of phthisis in the same family in which there seemed to be little doubt that each case in succession was an instance of contagion occurring in persons whose respiratory organs furnished a favorable soil. The second case occurred in a man who had, just previous to the contagion, passed a most searching examination when applying for insurance in the Mutual Life of New York. Yours truly,

W. L. MUNRO.

A note of inquiry sent in regard to the cases spoken of above was replied to as follows : —

111.

212 No. Main St., Providence, February 20, 1890.

Dr. H. C. Ernst.

Dear Sir, — Yours of the 14th inst. at hand. The particulars of the cases of which I spoke were as follows, as nearly
as I can give them just now. Amy, wife of Dr. W. W——,
of New York, returned to this country after several years'
residence in Europe, about October, 1886, in an advanced
phthisical condition. Her father and mother are living at an
advanced age. No deaths from phthisis in immediate family. Tall, thin, narrow-chested, and general phthisical aspect.
Resided here mostly with her brother-in-law, A. II. A——.
Inhabited same rooms as rest of family, no attempt at quarantine being made. Expectoration moist and abundant, generally received in an open cup, not disinfected. Whole house
kept habitually at a temperature of 80 F. or over. Died in
March, 1887. A. H. A——, similar physique, family history fair, passed a careful examination for life insurance for
a large sum in Mutual Life, about 1885 or 1886. Never
rugged, but very seldom sick. Shortly after Mrs. W——'s
death began to show symptoms (never correctly interpreted
by his then attending physician). By July 4th had marked
cough, with expectoration ; unable to attend steadily to busi-
ness. Ran through all of the stages, symptoms developing
somewhat slowly, and died March, 1889, having been practically bed-ridden for six months. Mary N. A——, wife of
preceding, and sister of A. W—— ; same build, intellectual,
but of highly excitable nervous organization, with a decided
tendency to hysteria ; was constantly with her sister ; slept
with her husband until September, 1888, about six months
before his death ; cared for him throughout. About September, 1888, developed a slight, persistent cough. Examination
of chest showed process already active. Progressed during
winter and spring. From June to September, 1889, was away
inland. Returned saying she had been perfectly well during
summer. Physical examination showed a rapid advance of

the disease, however. While laying plans for a winter in Colorado, her little girl was taken sick with typhoid. Before her convalescence, Mrs. A—— was herself taken sick, had a two weeks run of fever of moderate severity, followed by two exceedingly severe relapses of two weeks' duration. During the existence of typhoid symptoms, cough and expectoration disappeared, and no râles were heard. Period of convalescence ushered in as usual, but hectic supervened directly upon subsidence of typhoid fever; phthisical symptoms returned with redoubled force; patient never left bed, wasted rapidly, and died January 21, 1890. Amy A——, child of A—— and M——, six years old, slept with her mother for four months before being taken sick with typhoid. During fever had considerable cough, and moist râles were heard. Signs now negative (takes cold very easily), but if any lesion exists it is quiescent.

If above notes should be of sufficient interest, I can without much difficulty procure more accurate data in many directions, e. g., remote family history.

Very truly yours,   W. L. MUNRO, M. D.

A note of thanks was returned to Dr. Munro for the letter given above.

112.

While believing for many years that our milk supply might be a prominent factor in the dissemination of tuberculosis, I have never found an opportunity for demonstrating such a relationship.   ARTHUR H. NICHOLS.

55 MT. VERNON ST., February 7, 1890.

113.

ROCKPORT, MASS., February 10, 1890.

Have never seen a case from which I could wholly exclude contagion (personal), heredity, and the evils of moist and otherwise bad location. But have seen several, two in particular, in which I strongly believed the cause might be found in the milk of tuberculous cows.

Respectfully,   O. H. O'BRIEN, M. D.

A letter of inquiry was sent to Dr. O'Brien, asking for any details he could give in regard to the cases spoken of above. The answer received was as follows : —

114.

ROCKPORT, MASS., February 14, 1890.

DEAR SIR, — In reply to your favor of the 14th, current, in which you ask for details of two cases of tuberculosis whose cause, as I thought, might have been found in the milk of affected cows, I can only give the facts as I remember them after the lapse of more than fifteen years. Although on re-call they seem trifling and altogether inconclusive, I will state them as well as I can, and you may attach what weight you think proper to them.

1. Mr. J. G——, aged thirty-three years, family history good, no phthisis as far as I could discover, both parents liv ing in healthy old age, moved with wife and three children to a new location. Tenement near a fish wharf ; lot in front of house a rocky swamp. He was poor, and the family were ill-provided. He and his wife, soon after change of residence, sickened and died of phthisis, very acute in both cases.

2. Mr. de G——, a French West-Indian by birth, family history unknown, had lived in a house a few rods from the former, with the same accidents of surroundings, for sixteen years. Soon after the death of No. 1, this man took ill with pulmonary consumption, and died after a brief course of the complaint. Two daughters, fair and plump girls, ages about sixteen and nineteen, sickened and died soon after from the same disease. Now I learned that both these families had been using the milk obtained from the same cow. This cow had the reputation of being an extraordinary " good milker." I often saw the cow. She was poor as a cow could well be ; looked starved and sick, although I knew that she was well-fed. At the time I asked myself the question, was the milk the vehicle of tuberculous disease ? From these meagre data I was, and am of course, unable to answer it, at least affirma-

tively. All I can say is, that there appeared to be some grounds for such a suspicion. It is fair to add that two, I think three, deaths, occurred soon after (within four years) in the same near neighborhood, when the milk of the cow in question could not have been the cause. I am firmly persuaded that tuberculosis (phthisis) is a communicable disease in other ways, and perhaps more frequently, than by hereditary transmission. I remain, sir,

Yours very respectfully,   O. H. C. O'BRIEN.

Mr. HAROLD C. ERNST, HARVARD MEDICAL SCHOOL.

115.

SPENCER, Mass., February 15, 1890.

DEAR SIR, — I have never seen a case of tuberculosis which I could trace *directly* to a milk supply as a cause.

Yours,   E. W. NORWOOD.

A note of thanks was sent to Dr. O'Brien for his letter (114), and a request for any cases of suspicion on the part of Dr. Norwood received no reply.

116.

Have never seen a case where the connection was at all clear. In some cases have had strong suspicions.

Very truly,   J. C. D. PIGEON.

ROXBURY, February 8, 1890.

A letter asking for information in regard to the cases where milk was suspected as a cause of tuberculosis received no reply.

117.

PHILADELPHIA, February 10, 1890.

DEAR SIR, — No, but I do not feel that I have paid sufficiently close attention to this question in all cases under my observation to render my negative report of any statistical value.   Yours truly,   W. PEPPER.

118.

HAROLD C. ERNST, Esq., M. D.

DEAR DOCTOR, — I have never in all my practice seen a case of tuberculosis which I could trace to a milk supply from the cow or goat. I have given much consideration to the subject, and have tried to investigate it.

Very resp. and truly yours,

WM. H. PANCOAST, M. D.

1100 WALNUT ST., PHILA., February 9, 1890.

A note of thanks was sent to Dr. Pancoast in return for the above.

119.

CHICOPEE, MASS., March 12, 1890.

Dr. H. C. ERNST.

In answer to your question regarding tuberculosis, I cannot say that I have been able to trace any case to the milk. For the most part, in the country here, our milk is from healthy animals.

Yours very truly,

F. F. PARKER.

120.

DORCHESTER, February 8, 1890.

I have seen many cases of tuberculosis, but none, I think, where I could not trace a family history as a presumable cause rather than any milk supply. Still I have no doubt that milk from tuberculous cows could be infectious.

Respect. F. S. PARSONS, M. D.

120 a.

Dr. F. F. Patch (Boston, Mass.) writes, "Nothing approaching it."

121.

BOSTON, February 8, 1890.

DEAR DOCTOR, — Excuse me, I never use a *postal-card*. I have made many thousand microscopal examinations of milk within the last twenty-five years. I have had a large number of patients that have had tuberculosis, but have never been able to trace its cause to a milk supply, although I have been looking for a cause.   Yours truly,

A. F. PATTEE.

122.

WORCESTER, February 11, 1890.

I have never seen a case of tuberculosis which seemed to have any connection with milk supply. Worcester cows are all healthy!   C. A. PEABODY.

123.

LITTLETON, MASS., February 12, 1890.

DEAR SIR,— I have never been able to trace to a milk supply as a cause of tuberculosis. The cows are usually healthy and well fed, — hence good milk. We have but little tuberculosis in this town.   Yours truly,

R. H. PHELPS.

124.

Dr. H. C. ERNST.

DEAR SIR, — I cannot state positively that I have ever seen a case of tuberculosis which could be traced to a milk supply as a cause, with absolute certainty.

Very truly yours,   R. B. PRESCOTT, M. D.

NASHUA, N. H., February 8, 1890.

125.

A note asking for further information from Dr. Prescott was answered as follows : —

NASHUA, N. H., February, 1890.

DEAR DOCTOR, — The cases I had in mind when I answered your circular letter, occurred some fifteen or twenty years ago, when I was in general practice in New York city, and were dispensary cases. But it was so long ago, that the evidence, facts, and details are wholly gone from mind, and I am unable to recall anything which would be of service to you at the present time, which for your sake I very much regret.   Very truly yours,   R. B. Prescott.

A note of thanks was sent for the above.

126.

Baltimore, 611 No. Calvert St., February 10, 1890.

Dear Doctor, — I have no personal knowledge of any case of tuberculosis which was traceable to infected milk.

Very truly yours,   Geo. H. Rohé.

Dr. H. C. Ernst.

A letter to Dr. Rohé, asking if he had information of any suspected cases, received the following reply : —

127.

Baltimore, 611 No. Calvert St., February 17, 1890.

My dear Doctor, — Dr. Wray, the state veterinarian of Maryland, whose office is in this city, spoke a few days ago of a family owning a number of cows, a large proportion of which were tuberculous. The family were likewise tubercular. He was not ready to say that there was any connection between the tubercular herd and the consumptive family. Possibly more detailed information, which he would doubtless be glad to give, would enable you to decide whether this instance is of any use to you. Should I get on the track of any cases likely to be of any interest to you, I will bear it in mind and will write to you about them.

Very truly yours,   Geo. H. Rohé.

Dr. Harold C. Ernst.

Neither a circular letter sent to Dr. Wray personally, nor an appeal to Dr. Rohé to ask about the cases spoken of above, met with any response.

128.

N. Y., 37 W. 35TH ST., February 8, 1890.

DEAR DOCTOR, — It is difficult to answer your question categorically. I have certainly suspected at times that the milk supply was one of the causes of tuberculosis. I never have been able to satisfy myself that it, by itself, was the efficient cause. Yet I can well understand that it may be.

Yours,   BEVERLY ROBINSON.

A letter was sent to Dr. Robinson asking for any cases in which he had suspected milk as a cause of tuberculosis, and the following reply was received : —

129.

NEW YORK, 37 W. 35TH ST., February 22, 1890.

DEAR DOCTOR, — I am afraid I cannot reply to your question in any such way as to be satisfactory, or of any real value. I have no recorded histories, or any positive facts to relate.

I have merely remarked that children have lost flesh and strength at times, without assignable cause, and with a very clear hereditary history. In such instances when the parents were closely questioned, I have found occasionally that the children were fed almost exclusively on milk. Now, when the source of the milk supply was inquired into, it was discovered that it was from a locality where I had reason to suspect that there was little or no intelligent supervision of the cattle, and where from the poverty of the people and their bad hygienic surroundings I premised that there might be tuberculous cattle in the herds.

You perceive at once that this statement offers little to a scientific inquirer in the way of real acquisition. It is, however, the best I can send you. I feel like adding that I regret my postal, as it may have given rise to some misapprehension in regard to the sum of my knowledge. It is now my practice, however, in cases of bottle-fed infants, to use every possible precaution in regard to the cow or cows who furnish the milk supply to the child. I consider auscultation of the lungs of the cow by a well informed veterinary to be one of the best means at our command, of detecting tubercular disease in its incipient stage. If this examination be positive as to its revelations, I shall hereafter direct that another cow be called upon to furnish the milk, and the diseased cow be isolated or gotten rid of altogether. At a more advanced stage of tuberculous disease in the cow, and always as an additional test, I believe expert examination of the milk, in view of your researches, should be insisted on, to recognize, if possible, the presence of the bacillus tuberculosis.

Sincerely yours,   BEVERLY ROBINSON.

A note of thanks was sent to Dr. Robinson.

130.

CHATTANOOGA, TENN., February 6, 1890.

DEAR DOCTOR, — There are so many chances for error in the search you are making for proof that tuberculosis may be communicated through a milk supply, that, notwithstanding the full warrant for perfect faith in this factor, I fear you will not be able to find a case that will stand the severe test that must be given in the interest of medical truth; I mean, to exclude every other cause or carrier which might be present in the nursery, in the household, in the schoolroom, in the street, in the workshop, factory, church, theatre, etc., etc.; indeed, in a hundred other ways through which the b. tuberculosis may be carried and find lodgment in the human body. I have just completed examinations of tissues from two cows — the material, lungs, mammary glands and sputum having been sent to me for that purpose from Michigan — in which typical specimens of bacilli of tuberculosis were present. I shall look for the result of your proposed task with much interest.   Very truly yours,

JAS. E. REEVES.

A note of thanks was sent to Dr. Reeves for this communication.

131.

No, although I believe strongly in the transmission of tuberculosis by milk, yet I have never seen a case in which I could trace the disease directly to that cause.

P. G. ROBINSON, St. Louis.

131 a.

BOSTON CITY HOSPITAL, February 15, 1890.

Dr. H. C. ERNST, *Harvard Medical School.*

DEAR DOCTOR, — In reply to your circular letter, I would say that after inquiry amongst several of our more recent house officers, and also among the visiting staff, I cannot find that we have ever had a case of tuberculosis here which it seemed possible to trace to a milk supply as a cause.

Yours very truly,
G. H. M. ROWE, *Superintendent.*

A note of thanks was sent to Dr. Rowe for his note and the trouble he had taken.

132.

No. 28 EAST 38TH STREET, NEW YORK,
February 11, 1890.

MY DEAR DR. ERNST,—I have your question regarding tuberculosis and milk supply. In reply I wish to say that I have never been able to trace a case of tuberculosis to this source. On the other hand, I have seen many cases when I could not in any way trace the cause of disease, and I await your conclusions with much interest.

Yours truly, NEWTON M. SHAFFER.

Thanks were sent to Dr. Shaffer for his note.

133.

February 10, 1890.

DEAR DOCTOR, — Have had no case of either phthisis pulmonalis, or tuberculosis, which I could have traced to milk as a cause (that is, bovine milk).

Yours, etc.,   E. L. SHURLEY.

A note was sent to Dr. Shurley asking for information in regard to any cases of transmission of tuberculosis by milk other than bovine which he might have seen, with the following reply : —

134.

25 WASHINGTON AVENUE, DETROIT, MICH.

Dr. HAROLD C. ERNST.

DEAR SIR, — I have seen two cases of miliary tuberculosis which I thought due to *mothers'* milk, but do not remember any due to the milk of other animals.

Yours sincerely,

E. L. SHURLEY,

(per S. E. S.)

135.

A request for fuller details of the cases mentioned by Dr. Shurley received the following reply : —

25 WASHINGTON AVENUE, DETROIT, MICH.,
February 25, 1890.

H. C. ERNST, M. D.

MY DEAR DOCTOR, — In regard to your request for the notice or observation of miliary tuberculosis by mothers' milk, I would say that I shall have to look up the cases in my note-book, and will do so as soon as I can, unless you are content with the fact that two children of healthy parentage, suffering from miliary tuberculosis, came under my observation, and who were nursed by tuberculous nurses. I will not enter into details unless you desire it, because in a month or two you will see some papers that will be published in the "American Journal of Medical Sciences" by Professor Gibbes and myself on the subject of tuberculosis and phthisis pulmonalis, in which I will appear as one who believes that miliary tuberculosis is not phthisis pulmonalis. Should you desire the details of the cases I will endeavor to look them up.

Yours truly,   E. L. SHURLEY.

A note was sent to Dr. Shurley asking for fuller details of the cases, at the same time suggesting that these might be some of the cases, details of which had already been refused by Dr. Gibbes (see his letters), and that therefore Dr. Shurley might desire to reconsider his decision to send the fuller details. This note was answered as follows : —

136.

HARPER HOSPITAL, DETROIT, MICH., March 1, 1890.

MY DEAR DOCTOR, — Your letter of the 29th is at hand. In reply would say that I am greatly interested in your work, and am ready to lend whatever aid I am able to; but of course must for the present accede to the desire of Dr. Gibbes — if he has any — in this matter, because he has been for the past two years working with me upon this subject. I have no doubt he will assent to my giving you such details as I have of the cases in question. I will consult him about it, and write you again. Concerning the vexed question of the nature of tuberculosis, which you incidentally mention in your letter, of course we cannot very well discuss it here. Suffice it to say that I too think that with a proper understanding we may not differ very much in our opinions.

Yours sincerely,   E. L. SHURLEY.

A note of thanks was returned to Dr. Shurley for this letter, but nothing has been heard from him in regard to fuller reports of the cases.

137.

I have seen a number of cases of intestinal tuberculosis in children fed on cow's milk, in which other causes could be safely excluded.   N. SENN.

A letter of inquiry to Dr. Senn asking for fuller details of these cases met with no response.

138.

3733 VINCENNES AVENUE, CHICAGO,
February 14, 1890.

MY DEAR DOCTOR ERNST, — In reply to your inquiry I would say that I have never seen a case of tuberculosis which I could possibly trace to impure milk. Theoretically I cannot accept such a cause of the disease possible.

Yours sincerely,

EDWARD WARREN SAWYER.

139.

I have never traced the cause of tuberculosis to milk supply in any case. I have seen infants nursed by mothers suffering from pulmonary tuberculosis without their (the infants) showing any immediate effects in the form of tuberculosis.

Dr. SKENE, Brooklyn.

A letter to Dr. Skene asking for details of the cases spoken of by him failed to receive a response.

140.

PHILADELPHIA, February 10, 1890.

MY DEAR DOCTOR, — I have not been able so far to get a *clear* indication that milk acted as a cause in any of my cases of tuberculosis. I will continue to watch closely and report to you any point in that direction that I may meet with.

Sincerely yours,   CHAS. E. SAJOUS.

A note to Dr. Sajous asking for any *suspicious* cases that he might have, received the following reply : —

141.

1632 CHESTNUT STREET, PHILADELPHIA,
April 22, 1890.

MY DEAR DOCTOR ERNST, — I did not answer your favor of the 14th of February, believing that I might be of some little use to you by inquiring as to the milk question among the patients (few in number) in my hands. So far nothing worth noting, negatively or positively, in any of them, has come up, the difficulty arising principally from their ignorance as to the source of their milk, etc.

Sincerely yours, SAJOUS.

A letter of thanks was sent to Dr. Sajous for his trouble.

142.

KANSAS CITY, Mo., February 13, 1890.

DEAR SIR, — Have never seen a case of tuberculosis that I could certainly trace to a milk supply as a cause.

E. W. SCHAUFFLER.

A note asking if Dr. Schauffler had had any cases in which he had suspected milk as a cause received no reply.

143.

BUFFALO, N. Y., February 18, 1890.

MY DEAR DOCTOR, — Such cases I have suspected, but I do not feel like stating that I have seen one.

Yours very truly,
CHAS. G. STOCKTON.

144.

LAWRENCE, February 14, 1890.

HAROLD C. ERNST, M. D.

DEAR SIR,— I am greatly interested in the subject of your circular, and have paid considerable attention to it. I could not trace any case of tuberculosis to milk supply, but I have seen cases of tuberculosis in cows in the surrounding country which I feel sure would give rise to tuberculosis in a fit subject drinking such milk.

Sincerely yours,

ANDREW F. SHEA, M. D.

145.

WILLIAMSTOWN, MASS., February 10, 1890.

DEAR SIR,— In reply to your inquiry must answer no. The question, you know, is one that has only a very recent basis on which to ask it.   Yours truly,

R. M. SMITH.

146.

Suppose one person in a hundred uses the milk of tuberculous cows, and that one sixth of all persons die of tuberculosis ; one in six hundred die using that kind of milk ; now I should think it extremely hazardous to trace the certain relations of cause and effect in *any* case whatever. I know nothing about the effect of our digestive processes as destructive or preservative of bacillus or any such organism whatever.

Dr. JOHN SPARE, New Bedford.

146 a.

No ; the breast milk probably has nothing to do with these cases of hereditary notions.

W. E. SPARROW,

Mattapoisett, Mass.

February 7, 1890.

147.

I like your question because I cannot answer it. I am disgusted with most of the milk I find among my patients. I am very glad indeed you have in this manner called my attention to the milk supply question. I hope to hear from you again. I will study this subject.

Yours cordially,   GEO. E. STACKPOLE.

148.

I cannot now recall a case of tuberculosis in my own practice directly traceable to milk supply.

JAS. CAREY THOMAS.

28 MADISON AVENUE, BALTIMORE.

A letter asking for any cases that Dr. Thomas had heard of met with no reply.

149.

DEAR DOCTOR, — I believe I know of no case where I consider the connection directly traced.

JAMES K. THACHER,
New Haven.

A letter asking Dr. Thacher for an account of any cases where the suspicion had been aroused in his mind received no reply.

150.

DEAR DOCTOR, — I can recall no case of phthisis which I could attribute to the use of milk from tuberculous cows. It would be hard to do so, because most of my patients receive their milk from out-of-town dealers, the condition of whose cows is unknown to me.   Very truly yours,

H. C. TOWLE.

July 10, 1890.

151.

No; our cattle in this vicinity are remarkably free from disease.

G. J. TOWNSEND,

South Natick, Mass.

152.

SUNDERLAND, MASS., February 13, 1890.

Dr. H. C. ERNST.

DEAR SIR, — I have never seen a case of tuberculosis which I could trace to the milk supply. Your inquiry leads me to say that I have often wondered whether the common barnyard fowl ever communicates this disease. It is usually well cooked, to be sure. It is, however, the filthiest feeder of any food animal in common use amongst us, — human excrement, the sputa of phthisical persons, and the like vile foods being apparently as palatable to the ordinary hen as the choicest viands, while the opportunity of picking up such foods are ample and usually made the best of.

Yours truly,

C. G. TROW.

153.

DEAR Dr. ERNST, — I never have been able to trace a case of tuberculosis to lacteal origin; but then, I have never tried. I am glad that you are looking the matter up, but fear that the time is premature. Yours,

T. G. THOMAS.

A letter was sent to Dr. Thomas asking *why* he thought the "time premature"? He replied as follows: —

154.

600 MADISON AVENUE, NEW YORK.

MY DEAR DOCTOR, — What I meant was this, — that the subject is yet so young that time has not been afforded for testing the validity of the theory advanced. I am glad that you have entered upon the inquiry, for the question is one of the most important that could come up for investigation.

Sincerely yours,

T. Gaillard Thomas.

A note of thanks was sent to Dr. Thomas for his note.

155.

Cincinnati, February 21, 1890, 100 W. 8th St.

Dear Doctor Ernst, — I have had cases, one or two, children, of basilar meningitis secondary to intestinal affections and independent of bronchial catarrh, in new houses, parents and attendants unaffected, brought up on the bottle, which I could interpret in no other way, especially as the milk used was from one cow only.   Yours truly,

J. T. Whittaker.

A note was sent to Dr. Whittaker, asking if any of the cows from which the milk spoken of in the above note were proven to be affected with tuberculosis, with the following reply: —

156.

Cincinnati, February 28, 1890.

My dear Doctor, — The cows in both cases were apparently healthy. No examination was made of the milk. I mentioned the cases because I could find no other explanation for origin. The houses were new, the parents and attendants free from all signs of the disease, and the surroundings (rural) perfectly good. The disease had not existed in even remote ancestry. I say this for the benefit of believers in heredity, of which I am not one. But the milk was taken from one cow in each case, and intestinal catarrh was the forerunner of the meningitis.   Sincerely yours,

J. T. Whittaker.

A note of thanks was sent to Dr. Whittaker for the above note.

157.

1413 WASHINGTON ST., BOSTON, February 11, '00.

MY DEAR SIR, — In reply to your circular, — my field of observation has been a very large one. I would not assert that milk was the direct cause in the adult consumptive (tuberculous). It may be quite different with children. I might add that the query is in its infancy, and difficult to solve for the present. P. D. WALSH.

158.

DEAR SIR, — I do not think of any case of tuberculosis that I could trace to milk supply as a cause. I have no doubt but tuberculosis in man may come from the consumption of milk from diseased cows. The sale of beef, such as I have seen, should be punished with death.

Yours truly, R. C. WARD, M. D.

NORTHFIELD, MASS., February 8, 1890.

159.

DEAR DOCTOR, — Please excuse my delay in answering your circular, which has been almost forgotten, because I have no statistics to offer. I have not seen a case of tuberculosis which I could trace to a milk supply, but I think that the public should be protected against the use of the milk or the flesh of tuberculous animals.

Yours truly, J. R. WEBSTER.

17 DIX ST., DORCHESTER, February 24, 1890.

160.

No, never except maternal.

ROBERT WHITE, M. D.

The following note was sent to Dr. White : —

161.

BOSTON, March 15, 1890.

MY DEAR SIR, — In the reply that you sent me to my query in regard to the transmission of tuberculosis by milk, you say "never except maternal." Will you be kind enough to give me some account of any such cases that you have, the fuller the notes that you will send me the better? Of course names are not necessary. It will be a great help to me, if you can see your way to doing what I ask.

Very truly yours,   HAROLD C. ERNST.

Dr. ROBERT WHITE, BOSTON, MASS.

The following was Dr. White's reply : —

162.

BOSTON, March 17, 1890, 331 Hanover St.

Dr. ERNST.

DEAR SIR, — The expression "never except maternal" means that like produces like.   Very truly yours,

ROBERT WHITE, M. D.

Dr. White's reply was so little courteous that no answer was returned to it.

163.

POMEROY, IOWA, February 11, 1890.

Having always had a country practice, where as a rule milk is pure and plentiful, I have never seen a case of tuberculosis which seemed traceable to a milk supply as a cause.

Yours, etc.,   D. W. WIGHT.

164.

DEAR DOCTOR, — I think I have seen many such, for example, tubercular disease from milk, mostly in hand-fed babies of perfectly healthy parentage, developing tabes mesenterica, phthisis, tubercular meningitis, yet I cannot prove it scientifically in a single case.

EDW. T. WILLIAMS, M. D.

ROXBURY, February 8, 1890.

An appeal to Dr. Williams for any details he could give me received the following reply : —

165.

Roxbury, February 13, 1890.

DEAR DOCTOR ERNST, — Have you ever seen a monograph by a Dr. Klenke, of Leipsic, " Ueber die ansteckung und verbreitung der scröfel-krankheit bei menschen durch den genuss der kuh-milch," 16mo, Leipsic, 1846? It is cited by West, "Diseases of Infancy," London, 5th ed., 1865, page 504, near the end of his last chapter on phthisis, with some brief but judicious observations of his own on the same point. This subject of tuberculous infection from milk has been in the air for fifty years at least. A vast deal has been talked and written about it. My own attention was very early called to it, and has been one of the motives of my long efforts to establish a diet kitchen in Roxbury, for the distribution of pure milk for sick children, and to help establish the Seashore Home, without thanks or profit to myself, but I think with substantial benefit to the community. I know that diseased milk breeds tuberculosis, but when you ask for details of cases, I am at a loss to give them. Details escape the memory, but leave behind impressions, and often convictions. My note-books are not indexed, and those of the Seashore Home inaccessible ; besides, my cases have been among the very poor and migrating sort of people, where it is difficult to get a complete family history, and you must eliminate heredity, or your case goes for nought. . . . My interest in abdominal tuberculosis was first excited by the unrivaled description of that disease in its enteric, mesenteric and peritoneal forms, by Rilliet and Barthez. There is no work in any other language, that I know of, that contains even a decent account of the disease. They show clearly that the old "tabes mesenterica," though illy named and described, is not a myth, but a reality. And my *impression* is formed on clinical, but not *post-mortem*, experience, that it is there, in the intestine and mesenteric glands, that we shall have to look for the earliest manifestations of tuberculosis from infected milk or cream. I am very sorry I can't aid you further, and wish you every success in your inestimably useful investigation.   Yours sincerely,

                                        EDW. T. WILLIAMS.

A note of thanks was sent to Dr. Williams for his letter.

167.

HAROLD C. ERNST, M. D.

MY DEAR SIR, — Your note of inquiry is at hand. The subject is of great importance and of especial interest to me, for, in consequence of tuberculosis, all extra pulmonary at first, I have instituted the very inquiry you suggest. While my suspicion was first directed to milk, then to food in general, I must retire without a single incident upon which to base a fact. I hope you will obtain something reliable. My difficulty has been with the general statements of the patients; in fact, after discovering that none of them have been large milk-drinkers, I conclude it would prove nothing if they were, for one drop of tubercularized milk would do the infection if subjective conditions were right. Of course we have facts concerning the possibility of producing general tuberculosis in animals. I believe that milk is a convenient vehicle, and the most probable one for human infection. Investigations into the sudden sicknesses of healthy infants, pointing to gastro-enteric irritation with subjective cerebral symptoms, would seem to me to be the field most likely to lead to positive results. . . . I await with great interest the result of your labors.   Yours very truly,

                                        HERBERT F. WILLIAMS.

BROOKLYN, N. Y.

168.

225 MARLBORO' STREET, February 8, 1890.

MY DEAR DOCTOR, — I have never seen a case of tuberculosis which I thought was caused by milk, although it is my custom to inquire upon this point. Very truly yours,

HAROLD WILLIAMS.

. . . .

Although receiving many more affirmative answers to the circular from veterinarians than from physicians, the correspondence was hardly as satisfactory. The letters that were thought proper to preserve for any reason are given below in alphabetical order.

169.

Dr. Bland (Waterbury, Conn.) speaks of a case occurring in his own family, — he having lost one of his own children with a suspected milk as the origin of the disease.

He was, however, unable, upon appeal, to trace the milk supplied to his family to a tuberculous cow, for the reason that his dealer furnished mixed milk. A portion of his letter follows : —

WATERBURY, CONN., March 22, 1890.

HAROLD C. ERNST.

DEAR SIR, — It seems that the milk-dealer who supplied my family does not produce from his own cows one fifth of the milk that he sells, but buys of other farmers. A few years ago there was a case of tuberculosis in a cow belonging to a farmer living about half a mile from this milk-dealer, and that is the only case that I have known in that neighborhood. And that was two or three years before I lost my child. Very truly yours,

THOMAS BLAND.

170.

Dr. ERNST, *Boston.*

DEAR SIR, — In my practice I do not get an opportunity often to observe the effect of milk from diseased cows, and so cannot recall a case to the point. Last week I examined a cow that died from tubercular mammitis, and that had been milked till within a week or so of death. If, as is generally believed, tuberculosis is contagious, then milk from such an udder must be particularly liable to transmit it.

Yours very truly,
J. WILLIAMSON BRYDEN.

171.

Dr. ERNST.

DEAR SIR, — I don't know just how I worded my note in answer to your query as to tuberculosis. What I meant to say was that I have known cases of cows that I believed were suffering from tuberculosis whose milk was distributed with other milk for family use, and that the deaths of children were frequently recorded as resulting from tuberculosis, making it possible that there was a relation between them and the milk supply. I have no data connecting any particular case with the milk supply from such cows. I have not kept my notes complete enough, were it possible to do so. I can give dates as to the cows, but not as to the children. A cow with tuberculosis, especially mammary, soon drops out, and passes from our observation and knowledge, — they are not apt to last long anywhere.

Yours respectfully,
O. H. FLAGG, V. S.

NEW BEDFORD, February 25, 1890.

172.

202 OLIVE STREET, ST. LOUIS, Mo., March 8, 1890.
HAROLD C. ERNST, Esq.

DEAR SIR, — . . . I believe that tuberculosis in cattle is a decided menace to the public health, — especially when there exist tubercular masses in the udder. . . .

Yours respectfully, H. F. JAMES.

173.

Mr. ERNST.

DEAR SIR, — . . . In regard to your first question, — as to whether I have known the disease transmitted from animals to the human subject, I cannot speak with any authority. As a matter of opinion my convictions are that it not only may, but that it does take place. I have known several cases of fatal " infantile diarrhœa," and one case of what the doctors called tubercles on the brain, occurring in a district where the disease was common among cows. These cases are culled from memory, and except in the case of tubercle of the brain I cannot say much definite about them. The latter, however, I do mind more about. He was a boy, five or six years old, — the son of a landed proprietor close to my native place. His father was much interested in rearing fine horses and besides kept a herd of standard grade of Ayrshire cows. I do not know that any of them were diseased, but they were fine-bred and in a district where tuberculosis was common.

Your second question is as to transmission from animal to animal. By the milk, I have not looked sufficiently close to speak with certainty. Have frequently had cases in young calves commencing with apparently ordinary diarrhœa, but which by and by became of a dysenteric nature and proved fatal in spite of remedial measures. These have occurred on places where tuberculosis existed, and in my own mind were set down as being tubercular enteritis. That animals do take it by ingestion of substances other than milk I am absolutely certain, from the fact that I have several times seen cases in older animals where there were bowel, mesenteric, and omental lesions, without any pulmonary trouble whatever. Again I have often met with cases where there were both thoracic and abdominal lesions, but the latter showed evidence of so much priority as to leave no doubt in my mind that the pulmonary lesions were secondary. Such cases, it seems to me, must have been due to infection through the alimentary tract.

Yours very truly,

GEO. F. KINNELL.

A note of thanks was sent to Dr. Kinnell for his letter.

174.

BINGHAMTON, N. Y., February 9, 1890.

Dr. H. C. ERNST.

DEAR DOCTOR, — Your circular received, asking for information in regard to the propagation of tuberculosis from milk, and I think I can give you the required information. Was called to see a herd of registered Jerseys that had been ailing some time; found herd affected with tuberculosis, and among them were three cows that had just dropped calves. Two of the cows were apparently in a healthy condition with the exception of being in an extremely emaciated condition and a large glandular enlargement of the mamma. These calves remained in a healthy condition until they were three weeks old, when they commenced to have diarrhœa which repelled all treatment, and finally died, one in four days and the other in one week. Diagnosed these as cases of phthisis abdominalis. In about one month the cows began to show symptoms of pulmonary trouble, and upon post-mortem found well-marked cases of phthisis pulmonalis, and the glandular enlargement was undoubtedly of a tubercular nature, but could not be certain, as I was situated so that a microscopical examination could not be made; but they having well-marked symptoms in the lungs should make it safe to assert that it was a tubercular deposit and caused the death of the calves. The other calf died of the same trouble in about three weeks, but the mother had no lesions in the mammary gland that could be discovered, but had large tubercular deposits in the mesentery and also in the lungs. If these cases will be of any value to you, I shall feel amply repaid, and if I can be of any service to you in the future I will be pleased to do it.

Yours truly,   G. A. LATHROP, V. S.

175.

NEW YORK, February 15, 1890.

DEAR SIR, — My positive transmission of tuberculosis has been experimental, having fed two rabbits and two guinea-pigs with such material, and developing tuberculosis; I have met in my practice here and there cases of transmission from cow to calf, fed with tuberculous material. The cases of suspicion of the transmission from animals to man can be obtained more from Dr. C. Peabody,[1] a veterinary practitioner of Providence, R. I.   Yours,

A. LIAUTARD.

176.

W. D. Middleton, V. S., sent a very interesting account of two children fed on the milk of tuberculous cows, both dying in from seven to nine months after the beginning of such feeding. His letter is too long for reproduction here.

177.

COLUMBIA, Mo., February 13, 1890.

HAROLD C. ERNST, M. D., *Boston, Mass.*

DEAR SIR, — Replying to yours of January, have to say that I have seen three cases of tuberculosis in human beings that seemed to have originated in cows' milk. I have positively induced tuberculosis in animals in five or six cases by

---

[1] See letter from Dr. Peabody (*infra*, 178).

feeding or inoculating milk from cows having tuberculosis in the udder.   Yours very respectfully,

P. PAQUIN,

State Veterinarian.

A request for further details in regard to these cases met with no response.

178.

PROVIDENCE, R. I., March 23, 1880.

FRIEND PETERS.

MY DEAR SIR, — Yours of the 21st received. I have not the dates of the case I reported, for at that time I was laughed at by many of the medical profession here, and I cannot now recall where I put the report of it, but I will give it you as near as I can remember. . . . I have found my note-book; it says: " June 15, 1878, — Mr. W—— called me to see a white and red cow, Ayrshire breed. Coughs, and is short of breath and wheezes. Pulse 60; respiration 14, and heavy at the flanks; temperature 104. Diminished resonance of right lung, but increased in part of the same. Emphysematous crackling over left lung, and dullness on percussion. Diagnosed a case of tuberculosis, and advised the destruction of the animal. December 12, — Cow in a cold rain a few days ago for about two hours. Milk still more diminished than at a visit made on September 25. Again advised the destruction of the cow. Family still using the milk. Respiration 20; pulse 85; temperature 104.6. February 22, 1879, — Temperature 104.8; respiration 26; pulse 68. Losing flesh fast. Milk still in small quantities. Advised as before to destroy the animal, and *not to use the milk*. May 30, — Called in a hurry to see the cow. Is now as poor as could be. No milk for a week. Pulse 80; respiration 40; temperature 106. The cow died in about three hours. Autopsy made fourteen hours after death; lungs infiltrated with tuberculous deposit. Weight of thoracic viscera 43.5 lbs.

Tuberculous deposits found in the mediastinum, in the muscular tissues, and in the mesentery, spleen, kidneys, udder, intestines, pleura, and one deposit on the tongue. The inside of the trachea was covered with small tubercles.

In August, 1879, the baby was taken sick, and died in about seven weeks. On post-mortem of the child there was found meningeal tuberculosis deposits all over the coverings of the brain, and some in the lung. In 1881, a child about three years old died with, as it was called, tuberculous bronchitis; and in 1886, a boy nine years old, who for three or four years had been delicate, died with consumption, " quick," as it was called.

So far as known, the family on both sides had never before had any trouble of the kind, and the parents were both rugged and healthy people, and so were the grandparents, one now being alive and 68 years old, and the other dead at 78.

Yours ever truly,

C. H. Peabody.

The above letter was sent to Dr. Peters, and by him incorporated in this report.

179.

Littleton, Mass., February 19, 1890.

Dr. H. C. Ernst.

Dear Sir, — I am in receipt of a letter from Dr. A. Worcester, Waltham, Mass., in which he wishes me to correspond further with you upon tuberculosis. The Doctor cites a case I told him of, but his memory is somewhat at variance with the facts. Tuberculosis I undoubtedly found in the barn, and have every reason to believe it was transmitted to the house, although I have not the *facts* to prove it. In the circular you sent me, the question — I took it — regarded the transmission of the disease to the human subject. If you mean transmitted to the bovines, I have proof sufficient to show that in one case, at least, I can prove it does so transmit.

If, as I said on my postal, the tubercular deposits in the mammary glands are sufficiently developed, I can see no reason why the disease should not be transmitted. If I can be of any further use to you on this subject, do not fail to call upon me freely, for I am delighted to be a co-worker with you in this important subject.

Yours very respectfully,
ALVORD H. ROSE.

Writing to Dr. Rose for further information in regard to the case of which he speaks in the above letter, he was good enough to send the following : —

180.

LITTLETON, MASS., February 22, 1890.

Dr. H. C. ERNST.

DEAR SIR, — Your letter of the 20th inst. is at hand. In reply I would beg to state that the case in question, of which I told Dr. Worcester, was indeed a suspicious case. In 1884 I was requested to examine a herd of Jersey cattle for contagious pleuro-pneumonia, in place of which I found tuberculosis. We had proof of the correctness of my diagnosis in the post-mortem examinations made. When I had finished explaining the effect that milk from a tuberculous cow would have upon a calf drinking it, they said the symptoms were identical with those presented by their grandchild the summer previous. This child had lived upon the milk of a cow that was known to have had tuberculosis for three years. Was this not a *suspicious* case? Let me give you a case near the point. Three years ago I visited a cow that had tuberculosis; she was filling the position of foster mother to an apparently healthy calf; she supplied milk to this calf for about 80 days, when the calf died from exhaustion, the result of obstinate diarrhœa. Post-mortem revealed the presence of tubercles in the mesentery, miliary tubercles on the costal-pleura, and two quite large tubercles in the inter-lobular tissue of the right lung. The mother of this calf died during parturition, and was said to have been a perfectly healthy cow, so we cannot say tuberculosis arose congenitally in this calf, but I think it a clear case of transmission of tuberculosis through the medium of the milk. There are several more cases I have seen that bear upon this question, but were I to give them they would be superfluous. Gerlach and others have produced tuberculosis by *ingestion*, in such animals as the pig, rabbit, monkey, chicken, and sometimes the dog. There is no question but that tuberculosis is the result of the presence of a specific bacillus; I am equally as certain that it is transmissible to man through the flesh and milk as it is from cow to cow through the milk. If I can be of any further use to you do not fail to call upon me. I am, sir,

Yours very respectfully,

ALVORD H. ROSE, D. V. S.

A note of thanks was returned to Dr. Rose for the above letter, and his consent was obtained for using it in any way that seemed proper.

A study of the preceding letters shows that from the medical men there came affirmative answers to the question asked as follows, and in these classes:—

From mother to child, 1 (Doggett, 51 *a*), 1 (Edwards, 53), 1 (Flagg, 59), 1 (Galvin, 69 & 70), 1 (Gordon, 74 & 75), 1 (Kollock, 90 *a*), 1 (Shurley, 133 to 136), or a total of 7. (This total means the number of gentlemen giving affirmative answers, *not* the number of cases they have seen.)

From cow's milk to child, 1 (Conant, 45), 1 (Deane, 51 *c*), 1 (Kite, 91 *b*), 1 (Gibbes, 61 to 64), 1 (Gage, 65 to 68), 1 (Hulbert, 85 *a*), 1 (Lovejoy, 99), 1 (Senn, 137), 1 (Whittaker, 155 & 156), 1 (Williams, 164 & 165), 1 (Bailey, 77 & 78), or a total of 11.

From animal to animal, 1 (Coates, 35 & 36); a total of one, (1).

Certain gentlemen expressed themselves as suspicious that they had seen such cases as were inquired about, viz. : —

1 (Bartlett, 9), 1 (Best, 13), 1 (Coffin, 44), 1 (Duer, 47), 1 (Dana, 49–51), 1 (Everett, 56), 1 (Hall, 83), 1 (Hull, 85 b), 1 (Hawes, 85), 1 (Prescott, 124), 1 (Leonard, 97), 1 (Mackenzie, 103 & 104), 1 (Macdonald, 105), 1 (O'Brien, 113 & 114), 1 (Pigeon, 116), 1 (Robinson, 128 & 129), — a total of 16.

Those expressing disbelief in such a source for the transmission of the disease are, —

1 (Dodge, 51 d), 1 (Garceau, 71), 1 Hutchinson, 85 d), 1 (Hunt, 85 c), 1 (King, 92), 1 (Mitivier, 108), 1 (Patch, 120 a), 1 (Sawyer, 138), 1 (Sparrow, 146 a), or a total of 9.

There were a number of gentlemen who said that they were out of practice or were specialists and had not seen cases of tuberculosis for years, — of these there were 15.

There were also others, not counted on any other list, saying that they had given no attention to the point whatever, or had not had it enter their minds; of these there were 61.

Of those making a simple negative reply there were 893.

There were received, then, of replies of some kind, —

Positive (mother to child),	8
(cow's milk to child),	11
(suspicious cases),	16
Negative (disbelief),	9
Negative simply,	893
Negative (out of practice),	15
Negative (no attention),	61
Total of replies to the letter,	1013

Percentages based upon such statistics as these are of the most misleading kind, and therefore no effort has been made to work out all that could be made ; but it seems reasonable to state one, — that showing the percentage of medical men whose attention has been attracted to cases such as the circular makes inquiry in regard to. In reaching this, it seems perfectly fair to deduct from the number of those to be considered, those who are out of practice, or who have not paid attention to the point. This would leave therefore 937 (1013 — 76) upon which to base the percentage. Counting all the positive and suspicious cases together, there are 35, and the percentage of those who have seen cases in which their suspicions have been aroused in this direction is 35 ÷ 937, or 3.7 —*per cent!* — a result that is as unexpected as it is surprising in its size, if one takes into consideration the difficulties surrounding the question, and the newness of the subject.

There are many other interesting things to be found by a careful perusal of the letters. The cases related by Dr. Munro (110 & 111) are interesting, although not coming within the exact scope of the question asked, and it is to be said that out of those who returned negative answers, and besides those already quoted in other ways, there were thirty that expressed their entire belief in the actual occurrence of such a method of transmission of the disease.

Letters of interest for various reasons, besides those specially referred to above, may be found in No. 15, by Dr. Blanchard, referring to the decrease of the disease in Sherborn; in No. 25, being encouragement from Dr. H. I. Bowditch; No. 38, from Dr. Chismire, referring to the prevalence of tuberculosis in Alaska; No. 52, from Dr. Edes, voicing the difficulties of the investigation; No. 55, from Dr. Page, giving interesting statistics in regard to the existence of tuberculosis among the insane; No. 58, giving Dr. Forchheimer's views upon the subject; Nos. 61 to 64, showing Dr. Gibbes' views; 75 *a* and *b*, giving Dr. Hodges' practice; No. 82, from Dr. Henry; 86, from Dr. Inches, giving his practice; 88 and 89, from Drs. Jenrett and Jacobi, emphasizing the difficulties of the query; 89 *a*, from Dr. Kinnicutt; 93, of encouragement from Dr. Lee; 118, from Dr. Pancoast, telling of his efforts in the same direction; 130, from Dr. Reeves, also emphasizing the difficulties in the way; 131 a, in which Dr. Rowe tells of his ill success in trying to find cases; 139, where Dr. Skene gives negative evidence; 145, where Dr. Smith emphasizes the newness of the question; 150, Dr. Towle also speaks of the difficulty of tracing the cause; 152, in which Dr. Trow suggests the common barn-yard fowl as a possible cause of the disease; 153 and 154, where Dr. Thomas speaks of the newness of the subject; 157, where Dr. Walsh says the same thing; and 167, in which Dr. Williams gives negative evidence.

So much for the correspondence from medical men. That from the veterinarians is much more positive, but for some reason it was much more difficult to obtain replies to letters of inquiry from them.

Of the replies received there were, — Positive, 1 (Clement, no *letter*), 1 (Culbert, no letter), 1 (Faville, no letter), 1 (Flagg, 171), 1 (Gardner, no letter), 1 (Huidekóper, no letter), 1 (Liautard, 175), 1 (Lathrop, 174), 1 (Middleton, 176), 1 (Paquin, 177), 1 (Peabody, 178), 1 (Rose, 179 & 180), 1 (Roberts, no letter), 1 (Trumbower, no letter), a total of 14.

Suspicious, — 1 (Kinnell, 173), 1 (W. P. Mayo, no letter), 1 (Bland, 169), 1 (Butler, no letter), 1 (Corlies, no letter), 1 (Howe, no letter), 1 (James, 172), 1 (Michener, no letter), 1 (Russell, no letter), — a total of 9.

There were sent in of negative answers 31. Therefore there were replies to the following extent: —

Positive,	14
Suspicious,	9
Negative,	31
Total,	54

This gives a percentage of persons among the veterinarians who have seen cases where they felt justified in suspecting such an origin of the disease as the investigation is seeking, of 23 ÷ 54, or 42.59 *plus* per cent!

Such a percentage is startling in its size, until one remembers the greater facilities that veterinarians have for observing such cases and their origin, when it does not seem so much out of the way, — granting that milk may be the vehicle for the disease that the experimental evidence offered in this paper tends to show that it is.

Combining the statistics obtained from the two sources, it appears that there were 991 answers received to the circular letter that should be counted, and that among these there were 58 gentlemen who have seen, or suspected, the existence of such cases as were inquired about, giving a percentage of 5.84 plus, which seems to be somewhat remarkable for the reasons already given.

Letters of especial interest, some of them having been already referred to, are 170, from Dr. Bryden, quoting a case; 178, by Dr. Peabody, especially.

As was said in speaking of the letters from the medical men, not a reply was received in which any suspicion of an expression occurred that could be twisted into meaning that the writer had seen or suspected such a case as was inquired about, but that at once a note requesting further information was dispatched. Many of these remained unanswered, but the original affirmative reply or suggestion was retained.

Finally, Dr. Jackson made a special inquiry in regard to tuberculosis among the Jews, and Dr. Peters one in regard to the prevalence of tuberculosis. The results of these lines of inquiry are given in the two reports here appended.

TUBERCULOSIS AMONG JEWS.

AMONG the replies received by Dr. Ernst, in response to a circular sent out in May, 1890, as to the frequency of tuberculosis arising from the use of milk of tuberculous animals, was a letter from Dr. Warriner, of Bridgeport, Conn., calling attention to an article in "The Nineteenth Century" for September, 1889. The article cited, " Diseases caught from Butchers' Meat," is by Dr. Behrend, of London. Dr. Behrend, after reviewing several articles proving the identity of bovine and human tuberculosis, speaks of the longevity of the Jewish race, and the comparative rarity of tuberculosis among this people; he draws the conclusion that the comparative rarity of tuberculosis in Jews may be explained, in part at least, by the inspection of all meat eaten by the Jews. After speaking of the hygienic laws of Moses in regard to the selection of meat for food, Dr. Behrend says [Quotation, p. 418] : " Finally the question . . . in the members of that faith."

Dr. Behrend's experience is similar to my own as Dispensary Physician at the North End of Boston.

I have made a careful review of all the cases of tuberculosis seen during my service as district physician. To my own cases I have added the cases seen by my predecessor and successor in this office, for several months, thereby obtaining statistics of this portion of the city for three years. The district includes Salem and all adjoining streets, and therefore takes in quite a large proportion of the poorer classes of the Boston Jews. Each case is entered but once on the books.

During three years, the following cases applied for treatment : —

5,937 Gentiles,   1 Jew = 5.1 Gentiles.
1,162 Jews.

Cases of tuberculosis: 196 Gentiles, 14 Jews. That is to say, among Gentiles, 1 case tuberculosis in 30.3. Among Jews, 1 case tuberculosis in 83, or tuberculosis was almost three times as frequent among Gentiles as among Jews.

At the same time, I would add that during these three years not a single case was entered, in 1,162 Jewish patients, of any disease directly or remotely dependent upon the abuse of alcoholic liquor.

Stalland, in a book on " London Pauperism," says: " Jewish children have no hereditary syphilis, and scarcely any scrofula." Casper Glatter writes [1864] : " Jews present remarkable immunity from intermittent fevers, convulsions, tabes mesenterica (abdominal tuberculosis), and from phlegmasiæ of the respiratory organs."

Jews attain a greater age than Gentiles, as proved by statistics throughout the world. In Prussia, in 1849, deaths were as follows : —

Evangelists . . . . 1 in 34.35 inhabitants.
Catholics . . . . . 1 in 30.18 "
Jews . . . . 1 in 40.69 "

Stalland gives the average life of Gentiles as 37 years; Jews, 49 years.

Statistics presented by Glatter to the Academy of Hungary, in 1856, give the average age of

Croats . . . . 20.2 years.
Germans . . . . 26 "
Jews . . . . 46 "

I have been unable to find any data as to the causes of death in Jews, but as tuberculosis causes so large a proportion of all deaths, in some crowded cities one fourth, or even more, of total deaths, it is reasonable to presume that a certain immunity from tuberculosis may be reasonably claimed as the cause of a part of the increased longevity of the Jews. Dr. Behrend, quoting from a paper by Dr. Noël Gueneau de

Mussy, "Etude sur l'hygiene de Moïse et des anciens Israel-
ites," gives a detailed account of the laws regulating the
choice of meat at the slaughter-houses; Dr. de Mussy gives
the details on the authority of the Grand Rabbi of France.
[Quotation, pp. 417, 418.]   "He (Moses) excludes . . . to
tuberculous lesions."   Dr. Burr, medical inspector at the
Brighton abattoir, says the Jewish butcher, often refuses a
large number of cattle, at times one third.   They refuse all
animals with any external injury.   In killing bullocks, their
law requires that the windpipe and half the œsophagus must
be severed; if more than half the œsophagus is cut through,
the carcass is refused.   In sheep and calves the whole œsophagus must be cut through.

If any pleuritic adhesions are present which are firmly attached to the lung, the carcass is refused.

To obtain some idea as to the sort of inspection made at
Brighton, I wrote to the Rev. Solomon Schindler, asking him
several questions, which I give with their answers.[1]

1st.   What is your personal experience as to the prevalence
of tuberculosis among Jews?

2d.   What were the dietary laws of Moses?

3d.   Are the liver and other entrails eaten?

4th.   What is the method of examination at Brighton?

Boston, October 1, 1890.

Dr. Henry Jackson.

Dear Sir, — In answer to your letter of September 28, I
shall follow closely the order of your questions.

1.   My personal experience is that consumption is as frequent among Hebrews as among other people.   Why should
it not?   The religious idea cannot prevent it or make a difference, and the dietary laws are hardly kept any longer, at
least not to the extent as they were formerly kept, or are still
kept, in Poland and Russia.

2.   The only good that came from the adherence to the

---

[1] Answer in letter.

dietary laws was that the post-mortem inspection of the animal proved whether it had been in a healthy condition at the time of death. The *lungs* were very carefully examined, and any sign of tuberculosis made the *whole* carcass prohibited meat. Controversies arose frequently between the authorized killer and examiner and the butcher, and tests were made to ascertain the effect of the tubercle upon the lung. If by blowing up the lungs air would escape, the animal was at once condemned.

3. Lungs, liver, milt, and tripe were not only allowed to be eaten, but were favorite dishes.

4. The present mode of examination is not reliable, and I have frequently advocated inspection by the city authorities, for the benefit of both Jew and Gentile. At present the wholesale butcher hires some Jewish cutter, — the class of which is fast dying out, — who kills and inspects the animal. He is paid by the piece, and receives an income only from what he declares *perfect*. The temptation lies near that he will close his eyes to many things. As a class, these people are poor and ignorant. They have merely learned the rules prescribed in the Talmud, and while symptoms may be absent of which they have a knowledge, the animal may perhaps be an unhealthy one.

Having declared a carcass fit for food, they attach their seal to it, and the retail butchers buy for their Jewish trade that class of meat. Supposing the wholesaler having sold out that kind of meat, and the retailer needing some, the temptation is near to take some other *healthy-looking* piece. The Jewish public care little about it; they trust in the butcher. The more orthodox classes, considering the eating of unallowed meat a sin, believe that the butcher who sells them the meat will be held responsible by God for the sin, and the liberals are too indifferent about it, and think that all meat that looks well is healthy, and allowed to be eaten.

I hope this explanation will suffice.

Yours respectfully,   S. Schindler.

The passage quoted from De Mussy, and the statement made by the Rev. S. Schindler, abundantly prove that the Mosaic laws, if carried out by a responsible person, even though uneducated, would, to a great extent, exclude the meat of tuberculous animals from the Jewish markets. Dr. Burr says that tuberculosis *per se* is not a sufficient cause for condemnation of the carcass by the Jewish butchers. Practically it is reasonable to suppose that the inspection in our slaughter-houses by Jews is a mere form.

Before drawing from such evidence any conclusion that the immunity to tuberculosis among Jews may be due to any care in the selection of their meat, it is necessary to consider what evidence has been obtained that the meat of tuberculous animals is infectious. In the first congress for the study of tuberculosis at Paris in 1888, one of the questions brought up for discussion was as to the danger arising from the use of the meat of tuberculous animals. Several observers reported results of experiments on animals. Nocard, Arloing, Galtier, all presented the results obtained from inoculating animals with juice squeezed from the meat of tuberculous animals. Nocard and Arloing especially expressed the opinion that though there is danger in the use of meat from tuberculous animals, the danger is slight. Nocard's experiments show that the bacilli of tuberculosis do not cause tuberculosis when injected into the blood current, and he succeeded in causing tuberculosis in animals only when the inoculations were made into the abdominal cavity. [These explanations are of much interest in view of the recent experiments as to the antiseptic properties of blood.]

Nocard concludes : —

" 1. Meat of tuberculous animals can under certain circumstances be dangerous.

" 2. But it is very exceptional that it is dangerous.

" 3. In such cases as it is dangerous it is so to a very slight extent."

Arloing, who made a few experiments, came to the same conclusions as Nocard.

Arloing inoculated ten guinea-pigs with portions of tuberculous organs. All became tuberculous.

Inoculated twenty guinea-pigs with juice of meat of tuberculous animals, and two became tuberculous. No mention is made as to the extent of the tuberculosis in the animals used for these inoculation experiments.

M. Butel, who offers no experiments to sustain the opinion he gives, considered that the meat of all tuberculous animals should be destroyed, no matter how advanced the tuberculous process was. He says: "Tuberculous meat and milk are a prominent, and perhaps the chief, cause of consumption in man." Butel concludes : —

"1. Is there danger in eating the meat of tuberculous animals ? Yes, it is the unanimous opinion of all scientists. 2. Is the danger great? It is formidable, both on account of the large number of tuberculous animals which enter into consumption, and the frightful number of persons that a single animal can infest, and finally, that each person in turn becomes an agent in the spread of the disease."

Cartier (inspector of abattoirs in Paris), says : —

"1. Tuberculosis is rare in calves, as attested by all veterinarians, and yet this animal is usually fed exclusively on milk.

"2. Tuberculosis is common in adult cattle, and yet they eat no meat.

"3. Why is it that in men as in cattle tuberculosis is usually pulmonary?

"It seems to me that if the infection were frequent from meat or milk the disease would attack especially the organs of the abdominal cavity, particularly the digestive organs."

After prolonged discussion as to the danger of using the meat of tuberculous animals, the following proposition was made by Chauveau, the president of the congress. The proposition was adopted almost unanimously, only three voting against it : —

"It is proposed to follow out by all means, including indemnity to owners, the general application of the principle of seizing and destroying all meat of tuberculous animals, no matter what the severity of the lesions in the animals affected." (Congrès pour l'Etude de la Tub. chez l'Homme et chez les Animaux. 1ère Session, 1888, p. 156.)

In the "Fortschritte d. Med." No. IV. 1890, is a review of an inaugural address by Kastner in Munich. Kastner inoculated sixteen guinea-pigs with the juice of the meat of twelve tuberculous cows. Some of the meat used came from animals condemned on account of generalized tuberculosis. The inoculations were made into the abdominal cavity; none of the guinea-pigs were tuberculous after two months. In the same number of the "Fortschritte" is an article by Steinheil. He used for inoculation the juice of muscle taken from human beings who had died of tuberculosis. Steinheil inoculated fifteen animals from juice of tuberculosis, and all the animals became tuberculous. Steinheil judges that his results differed from Kastner's in that the tuberculous process in the subjects used was so advanced, and hence draws the conclusion that the meat of animals with a very advanced form of tuberculosis might be dangerous.

It seems to the writer probable that the positive results obtained from the inoculations with meat of tuberculous animals are due to the presence of tuberculous glands in the tissue surrounding the muscular fibres, rather than tuberculosis of the muscular fibre themselves, a rare pathological condition. In advanced general tuberculosis many glands throughout the body are affected, and such carcasses are undoubtedly dangerous for use as food.

As yet no satisfactory scientific evidence has been offered that the meat of animals affected with localized tuberculosis is infectious.

While tuberculosis is rare in muscular tissue, it is the rule to find tuberculous lesions in the liver and other glands of the body, and that organ especially should never be used for food, no matter what the extent of the disease in the animals slaughtered. Meat that has been condemned should be destroyed, as the various methods used to preserve meat, as salting, drying, smoking, do not destroy many bacteria which are present. Direct experiment has shown that a tuberculous lung when salted was as infectious as before it was treated in that way.

HENRY JACKSON.

November, 1890.

UPON THE PREVALENCE OF BOVINE TUBERCULOSIS.

By Austin Peters, M. R. C. V. S.

Admitting that tuberculosis is due to a specific germ, the *bacillus of tuberculosis*, and that it can be communicated from one animal to another of the same or a different species, by means of the expectorations after they become dry, or by the consumption of the flesh and milk or dairy products of tuberculous cattle; yet, in order to appreciate the danger to human beings from the use of the dairy products of tuberculous cows, it is important to have some idea of its prevalence.

It is an impossibility to get any statistics to show the extent to which this malady exists among our bovine population, but I think I can show that it is of adequate frequency to be of very great importance from a sanitary and economic standpoint.

Fleming, in his "Manual of Veterinary Sanitary Science and Police," in speaking of the geographical distribution of this disease among animals, says: "Tubercular phthisis, or tuberculosis, probably prevails among the domesticated animals over the entire globe, though its frequency will depend upon various external influences, as well as the constitutional tendencies of different species and breeds. In some countries it is enzoötic and very destructive. Such is the case in densely populated districts and in unhealthy climates, or in regions where animals are improperly fed and housed. In Mexico, for instance, it is very common and causes much loss, — about 34% of the animals slaughtered for food being found affected. In Europe, particularly in the cow-sheds of the larger towns and cities, it is extensively prevalent; and in this country (meaning England) it has long been recognized as a common disorder among animals, but more especially as affecting the bovine species." Walley's "Four Bovine Scourges" considers contagious pleuro-pneumonia, rinderpest, foot and mouth disease, and tuberculosis, as the four great cattle plagues of the world.

In this country rinderpest is unknown; foot and mouth disease does not exist at present; contagious pleuro-pneumonia is confined to a limited area near New York city, it having been stamped out in every other locality in the United States where it has ever existed: so that, to-day, we can safely say that the only one of the four great bovine scourges staring us in the face, and challenging us to combat, if we are not afraid to grapple with it, is tuberculosis.

In France and Germany the regulations governing the veterinary inspection of abattoirs are very strict, and the inspections made there are the chief source of any figures upon the frequency of bovine tuberculosis to be obtained.

In 1887, the French government added tuberculosis to the list of contagious animal diseases for Algeria, and in 1888 classed it with the diseases recognized by the sanitary law of France. Consequently, at the present time, bovine tuberculosis is the object of repressive sanitary law in France and Algeria.

In France, every animal recognized as being tuberculous is isolated and sequestrated, and it cannot be removed except for slaughter, which is carried out under the surveillance of a sanitary veterinary surgeon. The consumption of the flesh of tuberculous animals is sometimes permitted under certain conditions ; that is, if the disease is slight and localized, the flesh is not considered dangerous; but if it is extensive and general, the carcass is condemned as unfit for human food.

In Germany, the practice is for the inspecting veterinarian at abattoirs to condemn carcasses of cattle suffering from general tuberculosis ; if the malady be localized, the carcass is marked in such a way that the consumer knows the animal was tuberculous, and the meat is sold at a reduced price, to be thoroughly cooked before being eaten.

Taking the statistics of the French abattoirs as a basis for arriving at results, M. Arloing, a French veterinarian, estimates that in France among the adult bovine population, five out of every 1000 are tuberculous.[1]

[1] *American Veterinary Review*, November, 1889.

According to the statistics of the Minister of Agriculture, there were on the farms of France on the 31st of December, 1887, 8,623,441 adult cattle. Including the cattle kept elsewhere than upon the farms the number would be about 9,000,000 as the adult bovine population of France. Admitting that the proportion of tuberculous is to the healthy as 5 to 1000, we see that the number of tuberculous animals is at least 45,000.

The mean value of these animals being estimated at 300 francs each ($60.00), the cost to the government for stamping it out, if done all at once, would be 9,000,000 francs ($1,800,000), or 6,750,000 francs ($1,350,000), depending upon whether they were appraised at two thirds or one half their value. This is not possible, as it would take several years to eradicate it, and the cost would be distributed over a considerable period, a little being expended at a time.

At the congress for the study of tuberculosis held in Paris, July, 1888, the following estimates were given as to the extent to which tuberculosis prevails among cattle : —

In England, according to Mr. Cope, the proportion is from 1% to 26%, depending upon the locality.

In Belgium, the proportion of tuberculous cows is estimated by M. Van Hersten as 4%.

In Holland, M. Thomassen reports the proportion of the tuberculous cattle to vary from 8.4 to 10.6 per 1000. At the abattoir of Augsberg, in 1887, the proportion of tuberculous cattle was 3.62%, and that of tuberculous calves was 0.013%.

Some of the German abattoir records [1] give us the following figures : —

Trapp reports that among 11,079 cattle killed at Strasburg abattoir in 1880, 220 or 1.9% were tuberculous (this number did not include those slightly affected). For the same year Mandel found 174 or 3.4% among 5,105 cattle slaughtered at the Mulhouse abattoir, and in 1879, Strobl and Magin recorded 1125 or 2.5% of 44,699 slaughtered at Munich. The

[1] Propagation of Tuberculosis by Lydtin, Fleming, and Van Hersten.

1125 tuberculous cattle killed at Munich that year were classified according to age as follows : —

Cattle under one year   2 or .2%
" from one to three years 81 or 7.1%
" from three to six years 378 or 33.5%
" over six years   664 or 59.2%

The 1125 tuberculous cattle may also be apportioned in the following order : —

218 or 1.13 % of 19,284 bullocks slaughtered.
558 or 5.3 % of 15,789 cows slaughtered.
40 or .68 % of 5,823 bulls slaughtered.
28 or .73 % of 3,803 young steers & heifers slaughtered.
1 or 0.0006% of 149,971 calves slaughtered.

From these figures we see that the disease is more common in cows over six years old than any other class of neat stock. This is due to the fact that, living longer, they have a longer time in which to acquire the disease, that their systems are depleted by giving immense quantities of milk, and that their hygienic surroundings are generally bad, they being kept in hot, badly ventilated, crowded, and often dirty stables, and deprived of the fresh air and healthful exercise accorded to other cattle. This is practically the case with the cows kept in the dairies surrounding large towns and cities, and it is among them that tuberculosis causes the greatest havoc and brings the percentage up, while the rest of the bovine population is comparatively free from it.

The statistics of the German abattoirs could be quoted until they filled a large volume, but a few suffice and more would be a mere repetition without adding to our knowledge.

The only American abattoir figures that I know of are some from the Brighton abattoir. Last year the Board of Health appointed as inspector at the Brighton abattoir a young veterinarian, Dr. Alexander Burr, a recent graduate of the Harvard Veterinary School, regardless of the protests and appeals of the practical politicians who wished to have the place, recently vacated by a former butcher, occupied by an impecunious cow dealer.

In a paper read at a meeting of the Massachusetts Veterinary Association, in June, Dr. Burr gives an account of his duties as inspector from October 1, 1889, to April 1, 1890; below I take the liberty of quoting his figures upon tuberculosis : —

Total number of cows and steers killed   15,506
Number of animals tuberculous, 28 cows, 1 ox   29 — .17%
Number of cows (eastern and western) killed 880 — 3.3%
Number of cows (eastern) 810, tuberculous   28 — 3.6%
Number of cows (western) 70, tuberculous   0 — 0.0%

That is, 29 animals were tuberculous out of 15,506, but one was an ox, the others were cows ; and these 28 tuberculous cows came from around this section of the country. I give the remainder of Dr. Burr's paper, together with the discussion which followed it, below : —

" Of the twenty-nine cases discovered, there was not one among them but showed pulmonary lesions. I do not wish to be understood as thinking there is no such thing as localized tuberculosis ; this has been demonstrated by inoculation, but from my experience it would seem that invasion most frequently takes place through the respiratory passages.

" Of course, we must take into consideration that the cows coming here are generally thought to be sound, that is, we do not get all the animals used in the cheaper grades of beef; thus it will be seen that the above statistics are not the actual statistics of the State ; still, I think, a fair average of abattoir statistics. An acquaintance with the subject of inspection, as reported in the current professional journals of the day, will convince any one, unprejudiced, that we are better off than any European country reported. That is, the percentage of tuberculosis among our animals is less than in any European country. The number of animals killed outside the abattoir can only be a very small number compared with the other, and it would be unfair to think that all such are diseased, or even one fourth of them.

"So far as can be judged from my short period of inspection, even among our eastern cattle, tuberculosis exists to a much less extent than among animals in the populous centres of most European countries, and among our western bullocks tuberculosis has almost no existence whatever, and this class of animals represent two thirds of our cattle population.

"I may add in connection with the foregoing, that in relation with the abattoir we have an establishment where fertilizers are manufactured, and dead animals of all kinds are received, such as horses and cattle, many of which are cows; these animals represent a fair average of the cows of our neighborhood; having died, the owners have seldom any disposition to hide them. I have examined all the cattle brought here and so far my record is as follows: —

Received dead cows at abattoir from October 1, '89,
    till April 1, '90,   80
Number found with tuberculous lesions,   6
Percentage,   7.5

"No better opportunity, it seems to me, could be found to reach a fair average of the extent to which the disease prevails among our animals."

The following discussion ensued: —

Dr. Howard stated that his personal experience with cattle was very limited, but hoped that Dr. Burr was right in his small estimation of the amount of bovine tuberculosis in the locality; he was afraid, however, that it existed to a greater extent than the essayist judged it to, from what some of our other practitioners say, in whom he has every reason to feel confidence.

Dr. Winslow's experience with tuberculosis was so limited that he had nothing to say about it.

Dr. Peterson thinks that a good many animals that are tuberculous are not sent to the abattoir; doubted if fifty per cent of the creatures with the disease were sent to the abattoir. He then told of a slaughter house out in the country, not a great way from Boston, which he happened to visit one day, and where he saw "strange sights."

Dr. Marshall said he thought there was less tuberculosis around eastern Massachusetts than many of our members would have us believe.

Dr. Stickney said he had but little cow practice, but he had seen a good deal of bovine tuberculosis. He thought that Dr. Burr's statistics were not very valuable towards showing the prevalence of the disease around here, as the beef he inspects comes chiefly from the West. Dr. Burr's statistics are only correct as far as the animals brought to the Brighton Abattoir are concerned, but do not prove a great deal beyond that. It is not to be wondered at that tuberculosis should exist in many of our well-bred dairy herds, as it has been carefully propagated there for years.

If Dr. Burr's figures upon the prevalence of bovine tuberculosis in this locality, as based upon the dead cows sent to the fertilizer manufactory, are correct, 7.5 per cent of the milch cows in the suburbs around Boston being tuberculous would be a rather alarming state of affairs; but when we consider that these figures simply apply to the cows sent to the abattoir, the estimate is more likely to be too small than too large. The N. Ward Company take a great many of the dead cows in the suburbs; the Muller Brothers, at Cambridge, take many more. We have no figures to tell us as to the condition of these animals when taken to these establishments. Then the dealers in cheap cows (and in fact more expensive ones for that matter) who attend the Watertown and Brighton markets know what a tuberculous cow is, although they may not know the disease as tuberculosis; they call such cows "coughers," and is it likely that they are knowingly going to sell a "cougher" to a butcher at the abattoir, or that an abattoir butcher is going to buy a "cougher" to kill when they know that there is a veterinary inspector ready to condemn the carcass as unfit for human food? No, the "coughers" are going to be sold to dealers in cheap beef, and bologna sausage manufacturers, whose slaughtering establishments are outside of the jurisdiction of the Boston Board of Health and safe from outside interference.

In the report of the Massachusetts Cattle Commissioners for 1888, is a special article by Dr. J. F. Winchester, of Lawrence, then a member of the Board, upon tuberculosis. He collected all the information he could upon the prevalence of tuberculosis in different portions of Massachusetts, by corresponding with the leading veterinarians all over the State, asking them to report the results of any inspections of herds which they made; many responded, myself among the number. Below I give Dr. Winchester's results as tabulated by him; the first table gives farms where the disease existed as confirmed by post-mortems upon some of the animals, the other gives a list of herds where the disease in all probability existed, although not confirmed by autopsy.

Herd.	Bovines on Farm.	Killed.	Suspicious.	Percentage killed.
No. 1	70	8	8	11.42
2	2	2	—	100.00
3	57	5	—	8.77
4	50	1	8	2.00
5	12	1	3	8.33
6	12	2	1	16.66
7	4	1	—	25.00
8	90	12	78	13.33
9	34	2	3	5.88
10	36	19	—	52.91
11	32	32	—	100.00
12	61	1	36[1]	1.65
13	14	8	—	57.14
14	5	2	3	40.00
15	4	4	—	100.00
16	7	2	5	28.57
17	30	4	2	13.33
18	5	4	1	80.00
19	25	7	2	28.01
20	35	6	—	17.18
21	2	1	—	50.00
22	1	1	—	100.00
23	1	1	—	100.00
24	8	3	—	37.67
25	28	4	—	14.28
26	30	4	—	13.33
27	44	30	14[2]	68.49
28	23	6	—	25.84
29	17	5	—	29.41
30	2	1	—	50.00
31	17	4	12	23.52
32	48	6	3	12.50
33	40	30	10[2]	75.00
34	20	20	—	100.00
	866	243	189	28 %

[1] Eleven otherwise disposed of.

[2] Disposed of otherwise.

Herd.	Bovines on Farm.	Symptoms of Disease, but none killed.	Suspicious.	Percentage that showed Symptoms of Disease, but none killed.
No. 1	24	2	—	8.33
2	13	2	—	15.38
3	12	1	—	8.33
4	8	—	—	—
5	38	1	2	2.63
6	15	2	1	13.33
7	11	2	6	18.18
8	7	—	—	—
9	30	2	3	6.66
10	28	2	3	7.14
11	15	1	—	6.66
12	11	2	2	18.88
13	12	6	2	50.00
14	3	2	1	66.00
15	17	3	4	17.66
	244	28	24	11+%

That is, in Massachusetts during 1887 and 1888, Dr. Winchester learned of 34 herds where tuberculosis actually existed as demonstrated by post-mortem examinations. The 34 herds contained 866 head of cattle, of which 243 or 28 per cent were killed as tuberculous, and 189 more were suspicious. In the 15 herds where tuberculosis in all probability existed, but where no post-mortems were obtained to prove it, there was a total of 244 head, of which 28 head or 11+ per cent showed symptoms of tuberculosis, and 24 more were suspicious. On the 49 farms there is a grand total of 1110 head of cattle, of which 271 are probably tuberculous, and 213 suspicious. Of the 213 suspicious, some were certainly tuberculous, and a number were disposed of in other ways than killing, that is, sold into other herds where the disease may not have before existed, to act as new foci of infection.

The following table of the cows owned by the Massachusetts Society for Promoting Agriculture, at their experiment farm at Mattapan, from January 1, 1888, to July 1, 1890, all of them being more or less tuberculous, is additional evidence of the frequency of this disease in Massachusetts, but one cow being obtained outside of the State.

List of cows owned by the Massachusetts Society for Promoting Agriculture, during experiments carried on at farm at Mattapan : —

Cow.	Where from.	Breed.
A	Peabody	Native
B	Milton	Jersey
C [1]		
D	Danvers	Native
E	Danvers	Native
F	Danvers	Grade Shorthorn
G	Danvers	Native
H	Danvers	Native
I	Danvers	Native
J	Peabody	Grade Shorthorn
K [2]		
L	Jamaica Plain	Native
M	Peabody	Grade Guernsey
N	Peabody	Guernsey
O	Newport, R. I.	Jersey
P	Framingham	Native
Q	Brookline	Guernsey
R	Brookline	Jersey
S	Jamaica Plain	Grade Guernsey
T	Wellesley	Jersey
U	Barre	Grade Guernsey
V	Cambridge	Native
W	Lynnfield	Jersey
X [3]		
Y	Brookline	Grade Ayrshire

It will be seen by the foregoing table that 22 tuberculous cows were used in the work, coming from 11 different towns, and representing 15 different herds. Of these, nine were natives, five Jerseys, two grade Shorthorns, two Guernseys, three grade Guernseys, and one grade Ayrshire ; the Channel Island cattle and their grades outnumbering any other class, the so-called native coming next.

In order to obtain still further information as to the prevalence and distribution of bovine tuberculosis, about 350 of the

[1] Bought for another purpose.
[2] Healthy, bought for another purpose.
[3] Showed no well-marked evidences of disease on post-mortem examination.

following circulars were sent out to veterinarians in various parts of the United States towards the end of the summer, with a blank to be filled out and returned to the sender: —

MASSACHUSETTS SOCIETY FOR PROMOTING AGRICULTURE,
23 COURT STREET, BOSTON, August 20, 1890.

DEAR DOCTOR, — I wish to collect some statistics to show the frequency, or infrequency, of tuberculosis among cattle in various parts of the country. If you will fill out the inclosed blank and return it as soon after September 1st as convenient, you will confer a great favor.

Yours truly,
AUSTIN PETERS, M. R. C. V. S.

P. S. More blanks will be furnished on application.

Seventy-nine answers were received to the circular, which may be classified as follows: —

Practitioners in large cities, whose practice is confined almost
exclusively to horses, hence they could report no cases of
bovine tuberculosis, 21
Veterinarians with a mixed practice, but had no cases in
the specified time, 19
Veterinarians reporting cases in their practice, 39
——
79

The following tables have been prepared from the answers of the two latter classes, those whose practice is confined to horses presenting nothing of special interest to tabulate.

It is preferable to number the reports, as some of the correspondents wish their names not to be made public; but I believe them to be reliable men, most of them being known to me personally or by reputation.

ANSWERS TO CIRCULARS SENT TO VETERINARIANS, SECOND IN ORDER ON CLASSIFIED LIST.

ANSWERS FROM VETERINARIANS HAVING MIXED PRACTICE, BUT REPORTING NO CASES OF TUBERCULOSIS.

Veterinarian.	Town.	Reply.
1	Roxbury, Mass.	No cases.
2	Newton, Mass.	No cases for two years.
3	New Bedford, Mass.	No cases in four and one half years' practice.
4	Fall River, Mass.	No cases.
5	Holyoke, Mass.	No cases. Thinks it decreases in Western Massachusetts.
6	Falmouth, Mass.	Has seen no cases in a six months' practice.
7	Providence, R. I.	Has seen no cases around Providence.
8	New Haven, Conn.	Has seen no cases in a year's practice.
9	Jersey City, N. J.	Has very little cattle practice; no cases for two years.
10	Bethlehem, Pa.	Has never had a case in his practice.
11	Pittsburg, Pa.	His practice is chiefly among horses; believes it to exist in cow stables about city, but dairy inspector is an ignorant butcher appointed for political reasons.
12	Charleston, S. C.	Has seen no cases there.
13	Savannah, Ga.	Has never seen a case in three years in the South.
14	Mobile, Ala.	Has had no cases since 1888.
15	Rushville, Ind.	That part of Indiana is almost exempt.
16	Chicago.	Never saw a case in his Nebraska experiences.
17	Bloomington, Ill.	Is rare in that part of State.
18	St. Joseph, Mo.	Has seen none in three and one half years' practice.
19	St. Louis, Mo.	Is rare; if anything, it decreases.

This table shows that tuberculosis is rare in certain localities among cattle, particularly at the South. It would also appear from the two tables that in New England the mild climate of what is known as the " South Shore " is less favorable for its development than the more rigorous climate of Maine, New Hampshire, and Eastern Massachusetts. In justice to Maine it must be said that the disease is kept pretty well under there, the report coming from the State Veterinarian, and representing the whole State (see second table).

When reports come from different portions of the same State, one giving cases, and another saying he has none, it helps to prove the infectious character of the disease, showing how it spreads in one locality, while it does not exist among the cattle of another.

ANSWERS TO CIRCULARS FROM VETERINARIANS WHO REPORT CASES IN THEIR PRACTICE DURING THE PAST YEAR, OR KNOW OF CASES.

ANSWERS FROM VETERINARIANS REPORTING CASES OF BOVINE TUBERCULOSIS.

Veterinarian.	Town.	Number of Bovines on Farms.	Tuberculosis.			Breed.
			Diseased.	Suspicious.	Total.	
1[1]	Portland, Me.	8 herds, 39 bovines, } in two years.	10		10	4 Jerseys, 4 grade Jerseys, 1 Holstein, 1 native.
2	Nashua, N. H.	4 herds, 133 bovines.	10	5	15	All natives.
3	Rochester, N. H.	6 herds, 97 cattle.	69	8	77	Natives, Jersey grades, Holstein and Shorthorn grades.
4	Boston, Mass.	4 herds, 8 cattle.	4	1	5	3 Jerseys, 1 Holstein.
5	Lawrence, Mass.	7 herds, 63 cattle.	12	9	21	7 Jerseys, 4 grades; 1 grade Guernsey.
6[2] 7[3] 8[4] 9	Littleton, Mass. Hanson, Mass. Springfield, Mass. Hartford, Conn.	8 herds, 174 bovines.	11	19	30	4 Devons, 3 grades; 2 Jerseys, 2 Guernseys.
10	Norwalk, Conn.	3 herds, 117 cattle.	17	10	27	1 Jersey, grade; 16 Holsteins; 1 Durham, grade.
11	New Haven, Conn.	1 herd, 8 bovines.		1	1	Grade Holsteins.
12[5] 13[5]	New Britain, Conn. Jamaica, L. I.					

			138		136	
14	Most of these cases seen in Orange Co., N. Y.	80 herds, 1250 cattle.				All breeds, one not more than another.
15 [7]	New York city.					Jersey.
16	Kingston, N. Y.		1	1	2	Jersey.
17	Poughkeepsie.	6 herds, 268 animals.	30	16	46	8 Jerseys, 2 natives, 20 Guernseys and Jersey grades.
18 [8]	White Plains, N. Y.					
19	Red Bank, N. J.	2 herds, 6 animals.	2		2	1 Jersey, 1 grade.
20 [9]	Elizabeth, N. J.					
21 [10]	Meadville, Penn.					
22	Philadelphia, Pa.	7 herds, 38 animals.	11	8	19	8 Jerseys, 3 grades.
23 [11]	Philadelphia, Pa.					
24	Chestnut Hill, Philadelphia, Pa.	2 herds, 14 animals.	6		6	Jerseys.
25	Meadville, Pa.	2 herds, 2 animals.	1	1	2	1 Jersey, 1 grade Jersey.
26	Easton, Pa.	1 herd, 13 animals.	2	1	3	Natives.
27 [12]	Baltimore, Md.					
28	Baltimore, Md.	3 herds, 48 cattle.	3	1	4	2 Jerseys, 1 grade.
29	Terra Haute, Ind.	3 herds, 30 cattle.	8	3	11	4 Jerseys, 4 natives. Thinks it increases.
30	Portland, Ind.	2 herds, 9 cattle.	3		3	Shorthorns. Thinks it increases.
31 [13]	De Kalb, Ill.					
32	Sterling, Ill.	10 herds, 241 animals.	27	22	49	20 grade Shorthorns, 4 grade Jerseys, 3 natives.
33 [14]	Milwaukee, Wis.					
34 [15]	St. Paul, Minn.	3 herds in 5 years.	5		5	Not very prevalent.
35	Davenport, Iowa.	3 herds, 112 animals.	11	6	17	2 Shorthorns, 6 grade Shorthorns, 3 Jerseys.

ANSWERS TO CIRCULARS FROM VETERINARIANS (continued).

Veterinarians.	Town.	Number of Bovines on Farm.	Tuberculosis.			Breed.
			Diseased.	Suspicious.	Total.	
36[18]	Columbia, Mo.	4 towns, 160 in one herd.	167	127	294	Mostly Shorthorns.
37	Vermilion, So. Dak.	1 herd, 34 animals.	1		1	Native.
38[17] 39	Lincoln, Neb. Spokane Falls, Wash.	1 herd, 63 Herefords.		3	3	Came from either Wisconsin or Illinois. Are isolated.
39 Vet.	17 States.	Over 165 herds. Over 2000 animals.	549	242	791	

[1] State veterinarian. [2] Inspector at U. S. quarantine station, at liberty to report no cases. [3] Knows of but one case. [4] Gives no figures. It occurs in his practice, but less frequently than formerly. His clients weed it out as it appears. [5] Knows of herds where disease exists, but can give no figures. [6] Has kept no record, but has seen two or three cases in past year. [7] Saw two Jersey cows, in consultation. [8] Has met with but few cases; is going out of practice on account of health. [9] Is out of practice, but knows it to be very prevalent, especially among Jerseys. [10] Has cases every year, but has kept no records. [11] Sees a few cases in consultation every year. [12] Sees cases, but has sent his records to Dr. Salmon. [18] Reports tuberculosis "alarmingly prevalent," but gives no figures. [14] State veterinarian. [16] Has kept no record of cases, but sees it often in his practice; saw 3 Jerseys in one herd die of it in one night. [16] State veterinarian and his deputies. [17] Met a veterinarian who said he had met but 6 cases in 6 years in Nebraska, — all imported.

From this table it will be seen that in the practice of 39 veterinarians, representing 17 States, most of them reporting for one year only, there occurred 549 cases of tuberculosis, 242 suspicious cases, a total of 791, among 165 herds, containing in round numbers about 3000 animals. That is, in the herds where tuberculosis existed, about 18 per cent were diseased and over 8 per cent suspicious, a total of about 26 per cent.

From the foregoing pages it will be seen that bovine tuberculosis is quite a common disease, particularly among the dairy herds of the East, and that the time is not far distant when action must be taken to prevent its spread among cattle, as well as to protect consumers from the use of tuberculous beef and dairy products.

I have presented, in the preceding pages, the evidence that we have been able to collect upon the points in regard to which information seemed to be especially needed. This evidence is sufficient, it appears to me, to warrant certain definite conclusions, as follows : —

1. While the transmission of tuberculosis by milk is probably not the most important means by which the disease in propagated, it is something to be guarded against most carefully.

2. The possibility of milk from tuberculous udders containing the infectious element is undeniable.

3. With the evidence here presented, it is equally undeniable that milk from diseased cows with no appreciable lesion of the udder may, and not infrequently does, contain the bacillus of the disease.

4. Therefore all such milk should be condemned for food.

     Respectfully submitted,   HAROLD C. ERNST.